AWS

FROM ABSOLUTE BEGINNER TO EXPERT.

THE ULTIMATE STEP-BY-STEP GUIDE TO
UNDERSTANDING AND LEARNING AMAZON
WEB SERVICES EFFORTLESSLY.

JASON RICE

TABLE OF CONTENTS

TABLE OF CONTENTS 3

INTRODUCTION 7

CHAPTER ONE – THE BASICS 9

   CLIENT-SERVER TECHNOLOGY 9
      DATABASE SERVER 12
      CLIENT APPLICATION 13
      NETWORK 13
      CLIENT-SERVER PROTOCOLS 14
      TYPES OF SERVER 16
      ORB 22

   COMMUNICATION NETWORK 26
      TYPES OF NETWORKS ARE: 26
      VARIOUS TYPES OF NETWORK PROTOCOLS. 28
      WHAT IS HTTP? 33
      WHAT IS HTTPS? 34
      UNDERSTANDING DNS (DOMAIN NAME SERVICES) 43
      IP ADDRESSES AND DNS SERVERS 46
      FINDING YOUR IP ADDRESS 49

   DOMAIN NAMES 50
      CREATION OF A NEW DOMAIN NAME 55
      SOA RECORD 56
      PROCEDURES TO ACCESS YOUR WEB HOST MANAGER (WHM) FROM YOUR DOMAIN. 63
      INSTRUCTIONS TO TURN A URL INTO AN IP ADDRESS .65
      WHAT IS A URL? 65
      DEFINITION OF AN IP. 66

CHAPTER TWO - AMAZON WEB SERVICE 70

   AWS EC2 76

   AWS LIGHTSAIL 78

AWS LAMBDA 79

THE TOP FIVE ADVANTAGES OF AWS FOR BEGINNERS 82

USES OF AWS SERVICES 88

INTERACTING WITH AWS 88

THE MANAGEMENT CONSOLE 89

COMMAND-LINE INTERFACE... 90

SDKS 91

BLUEPRINTS... 91

CREATING AN AWS ACCOUNT 92

SIGNING UP.. 93

SIGNING IN 96

CREATING A KEY PAIR.. 97

MAKE A BILLING ALERT 101

CHAPTER THREE - DEVELOPING A VIRTUAL
INFRASTRUCTURE 104

UNDERSTANDING AND USING REST APIS 105

THE ANATOMY OF A REQUEST....................................... 106

JSON... 108

HTTP STATUS CODES AND ERROR MESSAGES 113

PROGRAMMING INTERFACE VERSIONS....................... 115

EC2... 115

LAUNCHING A VIRTUAL MACHINE 117

SELECTING THE OPERATING NETWORK 118

VIRTUAL APPLIANCES ON AWS 119

CHOOSING THE SIZE OF YOUR VIRTUAL MACHINE...... 121

Configuring Details, Storage, Firewall, And Tags 124

Connecting To Your Virtual Machine 128

Installing And Running Programming Physically 131

Monitoring And Debugging A Virtual Machine.132

Showing Logs From A Virtual Machine...............133

Examining The Load Of A Virtual Machine 134

Shutting Down A Virtual Machine 136

Changing The Size Of A Virtual Machine 139

Starting A Virtual Machine In Another Server Center 142

Allocating an open IP address................................144

Adding a Network Interface To A Virtual Machine 147

Associating an Additional Open IP address to Networking Interface 149

Chapter Four - Securing Your Network 151

What is VPC (Virtual Private Cloud), Subnet in AWS? 153

How Route Tables Work...159

What is Internet Gateway 159

What are AWS Firewalls?..160

AWS Web Application Firewall 161

AWS Network Firewalls...163

Understanding Security on AWS? 165

THE AWS SHARED SECURITY MODEL 166

AWS SECURITY FEATURES 166

WHY AWS SECURITY IS SO IMPORTANT 170

KEEPING YOUR SOFTWARE UPDATED 172

INSTALLING SECURITY UPDATES ON RUNNING VIRTUAL MACHINES 173

AWS IDENTITY AND ACCESS MANAGEMENT (IAM) 175

AUTHENTICATING AWS PROPERTIES WITH TASKS 176

CREATING A PRIVATE NETWORK IN THE CLOUD: AMAZON VIRTUAL PRIVATE CLOUD (VPC) 177

CHAPTER FIVE - STORING INFORMATION IN THE CLOUD 180

STORING YOUR ITEMS: S3 AND GLACIER, AMAZON CLOUDFRONT 180

AMAZON S3 182

HOW TO CREATE BUCKETS 183

UPLOADING YOUR FIRST ITEM TO A BUCKET 185

VIEWING TRANSFERRED OBJECTS 188

BACKING UP YOUR INFORMATION ON S3 WITH AWS CLI 189

AMAZON GLACIER. 191

CREATING A S3 BUCKET FOR THE UTILIZATION WITH GLACIER 192

AMAZON CLOUDFRONT 194

AWS CLOUDFORMATION 195

FINAL REMARKS 199

INTRODUCTION

Hello, and welcome to *"Amazon Web Service"* *your Ultimate Step-by-Step Guide to Understanding and Learning Amazon Web Services Effortlessly.* If you are one of those that created enthusiasm for cloud computing, then Amazon Web Service is the best technological platform to drift in. Every section of this Book manages some particular ideas on AWS service and give a stable point by point specialized synopsis of the AWS service.

But to be honest, it is beyond the realm of imagination to cover all services offered by AWS without experiencing a few basics. Along these lines, I will be concentrating on the services that will best assist you with beginning rapidly to prepare you in understanding most broadly utilized Amazon Web Services effortlessly.

Amazon Web Services is a saturated form of learning to understand and adapt. You need to have an understanding of some fundamentals before venturing into AWS such as:

- Basis of Client-server Technology: the connection between a client (your PC internet program) and the server (the machine getting your browser program demands)
- Familiarity with various sorts of network protocols, similar to HTTP in terms of web technology, and also a protected form called HTTPS.
- Knowledge of an IP address.
- Understanding DNS (domain name services) and how to transform a URL in form of an IP address

Having a full comprehension of the above-account ed basics will set you up for the advanced section of this book.

The advanced section of this book covers a great deal about working on Amazon Web Service, which begins with creating an account on AWS to running a virtual machine and AWS security and lots more.

CHAPTER ONE – THE BASICS

Welcome to the absolute beginner part of this book, this section covers essential fundamentals needed before I start with things you need to know to operate on Amazon Web Services.

CLIENT-SERVER TECHNOLOGY

Client-Server technology is a method for isolating the features of an application into at least two particular parts. Client-server portrays the connection between two PC programs in which one program, the client, makes a service demand from another program, the server, which satisfies the request. The client exhibits and controls information on the PC. The server goes about as a centralized computer to store and recover ensured information. It is a network-style in which every process on the network is either a client or a server. Servers are incredible PCs or process devoted to overseeing disk drives, printers (print servers), or network server.

Furthermore, Clients are workstations on which a user has the authorization to run applications. Clients depend on servers for properties, for example, files and devices.

Frequently clients and servers impart through a PC connect with unconnected hardware, yet the client and server can dwell on a similar framework. The machine is a host server that is running at least one server programs that share their properties with clients.

A client doesn't share its properties and yet demands content from a server. Clients, along these lines, start correspondence communication with the servers that are awaiting approaching requests.

The character of the client-server portrays the relationship of projects in an application. The server part gives a capacity or service to one or numerous clients who start their service demands.

Capacities, for example, such as trading email, Internet access, and database, are assembled dependent on the client-server model. For instance, a web program is a client program running on a client's PC that can get to data collected on a web server or the Internet. That program may, thus, forward the request to its program database client that sends a request to a server database on another PC to recover ledger data.

The client-server model has gotten one of the focal thoughts of network computing. Numerous business applications being composed today utilize the client-server model. In promoting, the term has been used to recognize disseminated computing by pocket-sized PCs from the "computing" robust incorporated centralized computer PCs.

Each example of client programming can send information requests to at least one server associated. This way, the servers can acknowledge these requests, process them, and return the mentioned data to the client. Even though this idea can be applied to an assortment of explanations behind various kinds of uses, the style remains in a general sense of equivalence.

A client-server model has the accompanying three particular segments, each concentrating on a specific activity:

• Database server
• Client application
• Network

Database Server

A server deals with the properties, for example, database, which effectively and ideally among different clients that at the same time demand the server for similar properties. The database server fundamentally focuses on the accompanying obligations.

• Managing a single database of data among numerous simultaneous clients.
• Controlling database entry and other security necessities.
• Protecting database of data with backup and recuperation highlights.

• Centrally upholding worldwide information worthiness leads to client applications.

Client Application

A client application is a piece of the framework that clients apply to collaborate with information. The client application in a client-server model spotlight on the accompanying activity:

- Showing an interface between the client and the property to finish the task.
- Managing report rationale, performing application rationale, and Validating information section.
- Managing the request traffic of accepting and sending data from the database server

Network

The third segment of a client-server framework is the network. The correspondence programming is the vehicle that transmits information between the clients and the server in the client-server framework. Both the client and the server run correspondence programming that permits them to talk over the network.

Merits and Demerits of The Client-Server Model

A significant favorable position of the client-server model is a result of its unified style, which makes it simpler to ensure information with access controls that are upheld by security approaches. Likewise, it doesn't make a difference if the clients and the server are based on the equivalent working framework since information is moved through client-server protocols that are stage skeptic.

A significant demerit of the client-server model is that if such a large number of client's demand information from the server simultaneously, it might get over-burden. Notwithstanding causing a situation known as "network clog," such a large number of requests may bring about a disavowal of service.

Client-server protocols

Clients usually speak with servers by utilizing the TCP/IP protocol suite. TCP is an association configured protocol, which implies an association is set up and kept up until the application programs at each end have completed with trading messages. It rules how to break application information into packets that network s can convey, responsible for sending packages to and acknowledging packets from the network layer, and it also oversees stream control and handles retransmission of distorted packets just as an affirmation of all packets that show up. In the Open Network Interconnection correspondence model, TCP covers portions of Layer 4, the Transport Layer, and parts of Layer 5, the Session Layer.

Conversely, IP is a connectionless protocol, which implies that there is no proceeding with the association between the endpoints that are conveying. Every packet that moves through the Internet is treated as a free unit of information with no connection to some other group of data.

The Distinction Between Client Server Computing and Peer to Peer Computing

The significant contrasts between client-server computing and peer to peer computing are as per the following:

- In client-server computing, a server is a focal hub that services numerous client hubs. Then again, in a peer to peer computing, the hubs altogether utilize their properties and speak with one another.

- Client-Server computing is accepted to be a subcategory of peer to peer computing.

- In client-server computing, the server is the one that speaks with different hubs. In peer to peer computing, all the hubs are equivalent and offer information with one another straightforwardly.

Types of Server

I. File Server

- File Servers are helpful in terms of sharing data over the network

- The client passes a request for document record over an interface to the account in server.

- The file server is known to be the most straightforward sort of information service utilized for trading messages over the network to locate the mentioned information.

- The file servers give access to the remote server processors. In the run of a software, shared information, databases and backups are collected on disk and optical storing gadgets that are overseen by the file server.

II. Database Server

- The client passes the SQL demands as messages to the database server; the consequence of each SQL order is returned over the network.

- The code, which forms the SQL demand and the information exists in a similar machine, and the server utilizes its preparing capacity to locate the mentioned information back to the client, rather than passing all the account s back to the client. This outcome in significantly more proficient utilization of the circulated handling power.

- Note that the application code dwells on the client; along these lines, you should think about writing code language for the client, or you can purchase a shrink wrap inquiry tool.

- The database servers give the establishment to choose supportive networks and provide an essential job in information warehousing.

III. Transaction Servers

The client can summon remote techniques or services that live on the server with a SQL database motor utilizing the exchange server.

- The network trade comprises of a single request/response. The SQL proclamations either totally succeeded or fail as a unit.

- With a transaction server, you make the client-server application by composing the code for both the client and the server segments.

- The client segment incorporates a Graphical User Interface (GUI), while the server part comprises of SQL exchanges against a database. These applications are called Online Transaction Processing.

- The OLTP are strategic applications that require less reaction time, usually about 1 to 3 seconds.

- The OLTP applications additionally require enduring powers over the security and credibility of the database.

IV. Groupware Servers

- It includes the administration of semi-organized data, for example, content, picture, mail, notice sheets, and the progression of work.

- This client-server framework places individuals in direct contact with others.

The best models are Lotus Notes and Microsoft Exchange.

• Specialized groupware programming can be based on a seller's canned configuration of client-server APIs. Much of the time, applications are made utilizing a scripting language and structure-based interfaces given by the seller. Presently, numerous groupware items use email as their standard informing middleware. Additionally, the Internet is rapidly turning into the middleware foundation of decision for groupware.

V. Object Application Servers.

- The client-server application is composed of a lot of discussing objects with an object server.
- The client object utilizes the Object Request Broker (ORB) to speak with the server objects.
- The ORB finds an occurrence of that object server class, conjures the mentioned

technique, and returns the outcomes to the client object.

- The server objects must help the simultaneousness and sharing angles. The ORB and another age of CORBA application server unite everything.
- The business ORB's that consent to the Object Management Group's CORBA standard incorporates the following such as Iona's DAIS, Java Soft's Java IDL, and Expersoft's Powerbroker.

VI. Web Application Server

Web application servers are another class of Internet programming. They consolidate the standard HTTP servers with server-side segment structures. Practically they are fundamentally the same as the object servers.

- This model of the client-server comprises of slim, versatile, and "widespread" clients that engage in discussion to the super-quick servers. Here the Web server returns archives when clients request them by name. The clients and the servers convey utilizing and protocol like protocol called HTTP, which characterizes a straightforward configuration of directions, parameters which are passed as strings, with no agreement for composed data.

- In the instance of Microsoft, the Multi-Tasking Staff disseminated the object server. Though on account of CORBA/Java, Enterprise JavaBeans has become the standard exchange of Web Application servers.
- Some of these servers additionally give the COM/CORBA spans.

ORB

This is an object request broker (ORB), which is a middleware technology that oversees correspondence and information trade between objects. ORB is the object transport. It lets the object straightforwardly make a request to - and get reactions from – different items found locally or remotely. The client doesn't know about the mechanism used to speak with, initiate, or store the server objects. A CORBA ORB gives a vibrant configuration of dispersed middleware services. An ORB is significantly more modern than elective types of client-server middleware, including the protocol Remote Procedure Calls (RPC's), Message-Oriented Middleware (MOM), database stored methodology, and cloud services. CORBA is the best client-server middleware at any point characterized.

ORBs advance interoperations of disseminated object frameworks since they empower clients to assemble frameworks by sorting out items from various vendors that speak with one another utilizing the ORB.

ORB technology advances the objective of object correspondence across the machine, programming, and merchant limits.

The elements of an ORB technology are:

- Interface definition
- Location and conceivable actuation of remote items
- Communication among clients and item

Duties of the ORB

- Providing Illusion of Locality: The ORB has to give the illusion of locality, as it were, to cause it to show up as though the object is local to the client, while justly, it might dwell in an alternate procedure or machine.
- Hide the Implementation Details: The next, specialized step advance toward object framework interoperations, which is the correspondence of object across stages. An ORB permits an object to conceal its execution subtleties from clients. This can incorporate programming language, working framework, and object area. Every one of these can be regarded as a "straightforwardness," and diverse ORB

advances may decide to help various transparencies, accordingly broadening the advantages of object direction across stages and correspondence channels.

There are numerous methods for actualizing the essential ORB idea; for instance, ORB capacities can be assembled into clients, can be independent procedures, or can be a piece of a working framework portion. These fundamental structure choices may be fixed in a single item, or there may be a scope of decisions left to the ORB implementer.

There Are Two Significant ORB Technologies:

- The Object Management Group's (OMG) Common Object Request Broker Style (CORBA).

- Microsoft's Component Object Model (COM).

Communication Network

In the realm of PCs, networking is the act of connecting at least two computing gadgets to share information. Networks are worked with a blend of PC equipment and PC programming. Networks can be sorted in a few distinct manners. One technique characterizes the kind of network as indicated by the geographic area it ranges. On the other hand, networks can likewise be characterized dependent on topology or on the types of protocols they support.

Prologue to Network Type

One approach to classifying the various types of PC network plans is by their extension or scale. For valid reasons, the networking industry alludes to almost every sort of structure as an area network.

Types of Networks Are:

- LAN - Local Area Network

- WLAN - Wireless Local Area Network LAN - Wide Area Network
- MAN - Metropolitan Area Network
- SAN - Storage Area Network, Network Area Network, Server Area Network, or in some cases Small Area Network
- PAN - Personal Area Network
- DAN – Desk Area Network

LAN - Local Area Network

A LAN interfaces network gadget over a generally short separation. A networked place of business, school, or home usually contains a separate LAN, however here and there, one structure will include a couple of little LANs, and sometimes a LAN will traverse a gathering of close-by structures. In TCP/IP networking, a LAN is regularly, however, not executed continuously as a single IP subnet. A LAN thus frequently associates with different LANs, and to the Internet or other WAN. Most local area networks are worked with generally modest equipment, for example, Ethernet links, network connectors, and centers. Remote LAN and other further developed LAN equipment choices likewise exist. The most widely recognized sort of local area network is an Ethernet LAN. The littlest home LAN can have precisely two PCs, while an enormous LAN can oblige a large number of PCs.

Metropolitan Area Network

MAN is a network crossing a physical area more significant than a LAN yet littler than a WAN, for example, a city. A Metropolitan Area Network is ordinarily possessed and operated by a single element, for example, an administrative body or large organization.

WAN - Wide Area Network

WAN traverses an enormous physical separation. The Internet is the biggest WAN, spreading over the earth.

Various Types of Network Protocols.

In the realm of technology, there are immense quantities of clients' speaking with various gadgets in various dialects. That additionally remembers numerous ways for which they transmit information alongside the diverse programming they actualize. Along these lines, imparting worldwide won't be conceivable if there were no fixed 'norms' that will oversee how the client conveys information just as how our gadgets treat that information.

I will be discussing more on "protocols," which are a set of instructions that help in administering how a specific technology will work for correspondence. It tends to be said that the protocols are advanced dialects actualized through networking calculations. There are various networks and network protocols that clients utilize while surfing.

There are different types of protocols that help a significant and merciful job in speaking with various gadgets over the network. These are:

- Transmission Control Protocol (TCP)
- Internet Protocol (IP)

- User Datagram Protocol (UDP)
- Post office Protocol (POP)
- Simple mail transport Protocol (SMTP)
- File Transfer Protocol (FTP)
- Hypertext Transfer Protocol (HTTP)
- Hypertext Transfer Protocol Secure (HTTPS)
-

Now, I will briefly discuss each of them:

- Transmission Control Protocol (TCP): TCP is a well-known correspondence protocol which is utilized for conveying over a network. It separates any message into a progression of packets that are sent from source to destination, and there it gets reassembled at the target.
- Internet Protocol (IP): IP is structured unequivocally as addressing the protocol, and for the most part, it is utilized with TCP. The IP addresses in packets help in routing them through various hubs in a network until it arrives at the goal network.

TCP/IP is the most well-known protocol interfacing with the networks.

- User Datagram Protocol (UDP): UDP is a substitute correspondence protocol to Transmission Control Protocol actualized basically for creating loss- bearing and low- inactivity between various applications.

- Post office Protocol (POP): POP3 is intended for accepting approaching E-mails.

- Simple mail transport Protocol (SMTP): SMTP is intended to send and circulate active E-Mail.

- File Transfer Protocol (FTP): FTP permits clients to move documents starting with one machine then onto the next. Types of account s may incorporate program account s, content documents, and archives, and so on.

- Hypertext Transfer Protocol (HTTP): HTTP is intended for moving a hypertext among at least two network s. HTML labels are utilized for making joins. These connections might be in any structure, like content or pictures. HTTP is structured on Client-server standards, which permit a client network for setting up an association with the server machine for making a request. The server recognizes the request started by the client and reacts accordingly.

- Hypertext Transfer Protocol Secure (HTTPS): HTTPS is truncated as Hypertext Transfer Protocol Secure, and it is a standard protocol to make sure communication occurs among two PCs, one utilizing the program and the other getting information from the webserver. HTTP is utilized for moving data between the client program (demand) and the webserver (reaction) in the hypertext design, aside from the moving of information, which is done in an encoded position. Thus, it very well may be said that https defeat hackers from understanding or alteration of information all through the exchange of packets.

What Is A Protocol?

A Protocol is a set of laws that we use for explicit purposes. In the present situation, when we are discussing protocols, it is about correspondence - how we converse with one another. For example, a shoe seller talks in English, and because you are fluent in English, you can easily comprehend. Therefore, that makes English the protocol.

Presently, discussing the web, specifically, various protocols are utilized to impart. Essentially for end-clients, the most significant and obvious protocols are HTTP and HTTPS. Even though there are numerous different protocols too, HTTP and HTTPS protocols take into account the majority of the populace.

What is HTTP?

HTTP is a Hypertext transfer protocol. They are basically used for sending and accepting content-based messages. As we all know, PCs work in a language of 1's and 0's, which are known as binary language. Accordingly, possibly every configuration of 1's and 0's build something, which could be a word.

Suppose I need to compose 'a.' Presently, if 0 means 'a,' 1 means 'b,' and 01 means 'c', I can gather that a blend of 0's and 1's to build a word. Right now, content is now created and is being sent on the network. The PC deals with numerous commands – binary language, content, and some different configurations like byte codes. Here, what is being moved is content. I am underscoring on 'content' since the program deciphers this content, and the moment program interprets it, it becomes hypertext and the protocol that moves the content is alluded to as hypertext transfer protocol - HTTP.

Utilizing HTTP enables you to transfer a lot, such as videos, sounds, and pictures.

What is HTTPS?

Hypertext Transfer Protocol Secure (HTTPS) is the safe adaptation of HTTP, the protocol over which information is sent between your program and the website that you are associating. The 'S' letter at the end of the word HTTPS means 'Secure.' It implies all interchanges between your application and the website are encoded. HTTPS is recurrently applied to secure remarkably private online exchanges like web-based banking and online shopping request structures.

What is the significance of HTTPS?

We settled upon the way that what is being moved starting with one point then onto the next is content. To comprehend why HTTPS protocol, we initially should know how wi-fi switches work. Suppose you are at an air terminal and you are interfacing with the wi-fi, which is the property of an outsider. Presently, when you are conveying over HTTP, the content is being moved by their switch. What's more, on the off chance that I go to a low form of the internet switch, I can quickly check and read the content that is being transferred. Furthermore, for what reason do we need HTTPS when HTTP appears. Presently, to spare our information from such assaults, we have to encode that information.

Encryption and Encryption Levels

Encryption in a straightforward type of clarification is concealing data. There are different approaches to do as such. You probably heard these terms - 128 pieces encode HTTPS, and 64 pieces encode HTTPS. 128-piece Encrypt is a high encryption procedure, and it's hard to unscramble (interpret). On account of HTTPS, when the information is being moved on the wires, the man in the center may at present recognize what is being moved, yet unable to interpret, as a result of the information being encoded. Just the program will unscramble it and show it, and the server will decode it and use it for exchanges.

Choosing Between HTTP and HTTPS

If you are scanning for "How to introduce SSL Certificate," that search would be private to you, right? Regardless of whether you are perusing or searching for an item, scanning an object, for the most part, don't need others to think about it. As an end-client, I would need to keep it private. There are things I might not have any desire to stay private, and for those, I can utilize HTTP. Be that as it may, for individual data, banks, and organizational data, HTTPS has gotten a standard for them in terms of being secure.

What Else Would It Be Advisable for You to Think About HTTPS?

There is no denying the way that protection has an expense to it. There are two cons-

- HTTPS demands to set aside more effort to process.
- Because it needs more opportunity to process, it needs more hardware and server, which implies extra expense.

While for HTTP, you utilize lesser demands when contrasted with HTTPS as the correspondence happens quicker (without encryption and unscrambling). Notwithstanding, I won't allude to it as a constraint for HTTPS. It is exceptionally abstract and individual; therefore, I think of it as a minimal effort that we pay to guarantee our protection.

Building a protected network has been around for some time. Making a Secure network purpose is being driven by preferences of Google, Facebook, etc. as I had referenced this is fundamentally a result of the accompanying two reasons which are:

- User Data and User Privacy: Using HTTPS guarantees that you, as a designer, care about client information, client's protection, and security.
- Protecting Your Data: As a network expert/ developer, you would certainly never need to share your essential information with malicious members.

Essentials of IP Addresses Before Your Cloud Migration.

When preparing for a cloud movement, remember to prepare for IP address changes that could influence your outstanding burdens and how they connect with inward and outer network traffic. Cloud suppliers and data centers have a constrained pool of IP addresses that they possess, and they frequently re-utilize recently allocated IPs to augment them. You can't just move your current IP addresses alongside your services. Or maybe, you'll get a powerfully appointed inside and outer IP address.

In some rare cases, you could lose those powerfully relegated IPs on the off chance that you stop your cloud occasion (yet typically just in the event that you stop and deallocate the VM (Virtual Machine) properties — most suppliers will keep your IP allocated to you if your machine is delayed. Fortunately, there are a couple of approaches to keep IPs moderately static in the cloud.

At the point when you move your application, database, or web-server to a cloud, you'll need to reconfigure any associations that point to your old on-premise IP address — for instance, firewall designs or databases for applications that are facilitated on a different VM. This can become very tedious and, for sure awkward over several situations.

You ought to have the option to buy static IPs from your cloud supplier. While this is an extra expense included in your membership, you can keep a steady IP address for any open confronting services. Make sure to buy these static IPs before setting up and preparing your VMs so that you can allocate them in like manner. A static IP ought to sit before your cloud servers so that any approaching traffic will highlight your virtual server center. Approaching traffic is coordinated by your load balancers as needs are. You can configure single ports on each VM to guide traffic to explicit servers more readily.

A key advance is to refresh the DNS to mirror the new IP address inside the cloud, so approaching open traffic is highlighted as the cloud supplier instead of the past address. DNS servers can likewise work to reroute traffic to a powerfully active address, since approaching traffic is highlighting the domain name as opposed to a static IP.

There are additionally some outsider programming suppliers that permit you to pick your IP address and afterward relegate cloud endpoints. You can change those endpoints on the fly while holding your IP, so regardless of whether you move suppliers or your cloud supplier changes your inside IP address, your open one continues as before.

Cloud computing stages give virtualized computing services that can be quickly sent for creation. These services incorporate virtual machines made from predefined machine pictures, open static or transient IP addresses that could be powerfully relegated to virtual machines, virtual networks with inward IP addresses, and directing tables. Lastly, security rules that range at least one virtual machine. A virtual machine is generally made on a physical computer in a particular geographic area. The district of a virtual machine influences the services and the IP subnets accessible to it. Creating a virtual machine (alluded to as an EC2 case on Amazon Web Services) infers joining the virtual machine in a default virtual private cloud network with a lot of predefined security rules (i.e., availability of ports) and routing tables. After an EC2 occasion is propelled, it is allocated a personal IP address just as a transient outer IP address, which is liable to change at whatever point the EC2 case is quit (halting an EC2 occurrence brings down or suspends the due hourly charges, yet doesn't lose the information or its setup).

The Importance of IP Addresses

Any individual who has been utilizing the web for any time allotment will have gone over IP addresses. These unique gatherings of numbers are used to recognize a PC associated with a network. These addresses are one of a kind to explicit PCs or littler networks on the web and are used to send information starting with one spot then onto the next. There are two types of IP addresses being used on the internet: IPv4 addresses and IPv6 addresses. IPv4 addresses look like four gatherings of numbers extending from 0-255 isolated by dots, while IPv6 addresses are longer, and colons isolate the assemblies. The extraordinary idea of an IP address implies that they are utilized for something beyond correspondence, as the address can likewise be utilized to recognize a server.

Understanding DNS (domain name services)

The web and the World Wide Web are wild boondocks that depend on PC and codes to discover and share information and data. One of the most key instruments of the web is the Domain Name System or DNS. (funny enough, numerous individuals think "DNS" means "Domain Name Server," it truly means "Domain Name System.") DNS is a protocol that has a set of measures for how PCs trade information on the web and numerous private networks, known as the TCP/IP protocol suite. Its purpose is fundamental, as it causes convert straightforward domain names like "howstuffworks.com" into an Internet Protocol (IP) address, for example, 70.42.251.42, that PCs use to distinguish each other on the network. To put it plainly, it Is a network of coordinating names with numbers.

PCs and other network gadgets on the web utilize an IP address to course your request to the website you're attempting to reach. This is like dialing a telephone number to associate with the individual you're trying to call. Because of DNS, however, you don't need to keep your address book of IP addresses. Preferably, you can easily interface through a domain name server, likewise called a DNS server or name server, which deals with an enormous database that maps domain names to IP addresses.

Regardless of whether you're getting to a website or sending an email, your PC utilizes a DNS server to look into the domain name you're attempting to access. The best possible term for this procedure should be termed as DNS name resolution, and you would state that the DNS server settles the domain name to the IP address.

Without DNS servers, the web would close down rapidly. Yet, how does your PC know what DNS server to utilize? Commonly, when you interface with your home network, web access supplier (ISP), or WIFI network, the modem or switch that appoints your PC's network address likewise sends some primary network design data to your PC or cell phone. That design incorporates at least one DNS server that the gadget should utilize while interpreting DNS names to IP addresses.

The Work of Domain Name Servers

IP Addresses and DNS Servers

You recently discovered that the essential task of a domain name server, or DNS server, is to define (interpret) a domain name into an IP address.

That seems like a straightforward undertaking, and it would be, aside from the accompanying points listed below:

- There are billions of IP addresses as of now being used, and most machines have an intelligible name too.
- DNS servers (in total) are preparing billions of requests over the web at some random time.
- Millions of individuals are including and changing domain names and IP addresses every day.

With such a significant amount to deal with, DNS servers depend on network effectiveness and web protocols. Some portion of the IP's adequacy is that each machine on a network has a novel IP address in both the IPV4 and IPV6 guidelines overseen by the Internet Assigned Numbers Authority (IANA).

Here are a few different ways to perceive an IP address:

- An IP address found in the IPV4 standard comprises four numbers isolated by three decimals, as in 70.74.251.52

- An IP address found in the IPV6 standard shall consist of eight hexadecimal numbers isolated by colons, Since IPV6 is as yet another standard, we'll focus on the more typical IPV4.

- Each number in an IPV4 number can be described as "octet" since it's a base-10 likeness an 8-digit base-2 (double) number utilized in directing network traffic. For instance, the octet composed as 42 represents 00101010. Every digit in the twofold number is the placeholder for a specific intensity of two from 2 to 27, perusing from the option to left. That implies that in 00101010, you have each of

21, 23, and 25. In this way, to get the base-10 comparable, simply include $21 + 23 + 25 = 2 + 8 + 32 = 42$.

- There are just 256 potential outcomes for the estimation of each octet: the numbers 0 through 255.

- The IANA assigns specific addresses and ranges as saved IP addresses, which implies they have a particular activity in IP. For instance, the IP address 127.0.0.1 is held to distinguish the PC you're now utilizing.

Where does your PC's IP address originate?

In case of inquiry about your desktop or PC, it likely arises from a Dynamic Host Configuration Protocol (DHCP) server on your network. The activity of a DHCP server is to ensure your PC has the IP address and other network setups it needs at whatever point you're on the web. Since this is "dynamic," the IP address for your PC will likely change now and again, for example, when you shut down your PC for a couple of days. As the client, you'll presumably never notice such an excess of occurrence.

Web servers and different PCs that need a steady purpose of connection utilize static IP addresses. This implies a similar IP address is continuously allocated to that network interface when it's on the web. To ensure that the interface consistently gets the same IP address, IP connects the address with the Media Access Control (MAC) address for that network interface. Each network interface, both wired and remote, has a one of a kind MAC address installed in it by the maker.

Finding Your Ip Address

Coming up next are tips on the most proficient method to discover your PC's IP address. Note that the address will change intermittently except if you've decided to utilize a static IP (uncommon for end-clients):

- Windows — Though you can navigate the user interface to discover your network interface settings, one brisk approach to discover your IP address is to open the Command Prompt application from Accessories and enter this direction: ***ipconfig*** Mac — Open your Network Preferences, click Network, be certain your present network association (with the green speck close to it) is chosen, click Advanced, and click the TCP/IP tab.

- Linux or UNIX — Open a terminal application, for example, XTERM or iTerm. At the direction brief, enter this order: ifconfig
- Smartphones utilizing WIFI — Look at your telephone's network settings.

DOMAIN NAMES

Individuals are not excessively great at recalling series of numbers. We are acceptable at recalling words, in any case, and that is the place domain names come in. You likely have several domain names put away in your mind.

In a domain name, each word and sets of dots you include before a top-level domain shows a level in the domain structure. Each level alludes to a server or a gathering of servers that deal with that domain level. The furthest word at the left corner in the domain name; for example, www is referred to as hostname. It determines the name of a particular machine (with a specific IP address) in a domain, regularly devoted to a particular reason. A given domain can contain a large number of hostnames insofar as they're all remarkable to that domain. (The "HTTP" part represents Hypertext Transfer Protocol and is the protocol by which the client sends data to the website that is visited. These days, you're bound to see "HTTPS," which is a sign the data is being sent by a secure protocol where the information is encrypted. This is particularly significant in case you're giving a credit or debit card number to a website.

Each domain has a domain name server dealing with its requests, and there is an individual or IT group keeping up the account s in that DNS server's database. No other database on the planet gets the same number of requests as DNS servers, and they handle each one of those inquiries while likewise preparing information refreshes from many individuals consistently. That is one of the most stunning pieces of DNS — it is conveyed all through the world on a vast number of machines, overseen by a great many individuals, but then it carries on like a single, incorporated database!

Since overseeing DNS appears such a difficult task, many people will, in general, leave it to the IT experts. In any case, by learning a tad about how DNS functions and how DNS servers are circulated over the web, you can oversee DNS with certainty. The principal thing to know is the resolution behind a DNS server on the network where it dwells.

A DNS server will have one or more of the following as its essential undertaking:

- Maintain a little database of domain names and IP addresses frequently utilized on its network and representative name objectives for every single other name to different DNS servers on the web.

- Paired IP along with hosts and sub-domains for which the DNS server has authority.

-

DNS servers that take control of the core assignment are usually overseen by your web access supplier (ISP). As referenced before, the ISP's DNS server is a piece of the network setup you obtain from Dynamic Host Configuration Protocol when you log unto the web. These servers dwell in your ISP's server centers, and they handle demands as follows:

- If it doesn't have the domain name and IP address in its database, it contacts another DNS server on the web. It might need to do this on various occasions.

- If it needs to contact another DNS server, it stores the query results temporarily so it can rapidly resolve consequent requests to a similar domain name.

- If it has no ability to discover the domain name after a sensible search, it restores itself incase of blunder, demonstrating that the name is invalid or doesn't exist.

- The second class of DNS servers referenced above is commonly connected with web, mail, and other web domain facilitating services.

Although it's conceivable to enter an IP address into a web program into the request to find a workable pace, it's much simpler to enter its domain name. Be that as it may, PCs, servers, and different hardware can't get a handle on domain names - they carefully depend on parallel identifiers. The DNS's activity, at that point, is to take domain names and make an interpretation of them into the IP addresses that permit machines to speak with each other. Each domain name has at any rate one IP address related to it.

Creation of a New Domain Name

At the point when you need to make another domain name, you have to do the following:

- Use the Whois database to locate an interesting domain name that isn't yet enlisted. There are a few destinations that offer free Whois database look. If the search comes up unoccupied, you realize the domain name is accessible.

- Register the domain name with an enlistment center. There is a lot of accounts to browse, and some offer unique costs for enrolling the COM, NET, and ORG variants of a domain simultaneously, for enlisting for at least two years, or for facilitating the domain with a similar organization.

- If you're facilitating the domain at an unexpected organization in comparison to your enlistment center, configure the account to point your domain name to the right hostname or IP address for your facilitating organization.

Utilizing the DNS servers from your enlistment center or facilitating organization implies that you have a stopped domain. This infers another person claims the PC equipment for the DNS servers, and your domain is simply part of that organization's more significant DNS configuration. On the other hand, in case you're energetic about facilitating your DNS, you can set up your server, either as a physical or virtual machine. Whichever DNS configuration you settle on, that DNS server (or gathering of servers) turns into the SOA for your domain.

SOA Record

For any web client, it's essential they have to enter a website's URL in the program to show up at their ideal landing page. They don't see that the PC sets up the association with an IP address. This is because of the Domain Name System (DNS) and its name purpose work. Here, the domain's name is assigned to the necessary number configuration. For the network to work, the name servers must have zone documents. Thus, these straightforward content documents contain various DNS account s that empower the DNS in any case.

As to the SOA account, if the DNS stage that you are utilizing (Windows, BIND, and so on..) is consistent with the RFC 1035, the structure of the SOA record will be the equivalent.

You can see the settings of the SOA record either by getting to the domain zone's properties and tapping on the "Beginning of Authority (SOA)" tab or by opening the zoning document itself utilizing a content manager (accepting that the zone is a standard-essential, not Active Directory Integrated).

How Does SOA Record Work?

The DNS is a decentralized, various leveled network: Name servers don't supply data to only any server on the web, however, to those situated in an apportioned zone. For this reason, the DNS server directs zone account s. These are straightforward content documents in which the entirety of a zone's DNS accounts are accounted for. To build up the various specialists, each zone file must contain an SOA account. SOA, in this manner, represents the Start of Authority. The account likewise gives data on multiple issues, for example, regardless of whether the addressed server is even answerable for the request.

The SOA record is particularly significant on account of server groups. Rather than bearing the load for all demands, the requests are conveyed among various gadgets. All together for the zone documents to stay current on all servers, a zone move must be performed usually. To accomplish this, the "slaves" (for example servers configured lower on the chain of command) synchronize their information with that of the "master" server and how the SOA record controls the zone transfer. For this reason, this kind of DNS account gets different types of data.

Account Makeup

DNS account s, for the most part, comprise a few fields. In this section, you'll discover all the important data. In contrast with different types, the SOA record has numerous fields:

- <name>: zone name
- <class>: network class
- <type>: account type
- <mname>: ace name
- <rname>: email address for the mindful head
- <serial>: the gradual sequential number that indicates the zone account adaptation
- <refresh>: time specification when a slave must demand the present ace rendition
- <retry>: time specification when a slave should play out a failed demand endeavor
- <expire>: time specification from which a slave doesn't discharge additional data without input from the master.
- <minimum>: time specification for to what extent data might be held in reserve.

Whether your SOA is elsewhere or on your network, you can expand and adjust your DNS settings to add sub-domains, divert email and dominate other services. This info is saved in a zone file on the DNS server. In case you're running your server, you'll presumably need to alter the zone file in a content tool physically. Numerous enlistment centers today have a web interface you can use to oversee DNS for your domain. Each new design you add is known as an account.

Here is the list of the most common types of account s you can configure for your DNS server:

- Host (A) — This is the elemental mapping of IP address to have a name, which is the fundamental component for any domain name.

- Canonical Name (CNAME) — This is just like a nickname for your domain. Anybody accessing that will be automatically synchronized to the server indicated in the A account .

- Mail Exchanger (MX) — This maps email traffic to a particular server. It could show another hostname or an IP address.
- Name Server (NS) — This encompasses the name server data for the zone. On the off chance that you configure this, your server will let other DNS servers realize that yours is an absolute power (SOA) for your domain while caching query information on your domain from other DNS servers around the globe.
- Start of Authority (SOA) — This is a significant account toward the beginning of each zone file with the essential name server for the zone and some other information. In a situation whereby a hosting company is running your DNS server, you won't have to deal with this.

View Your SOA record

SSH command dig would show the below:

dig SOA domain.com;; ANSWER SECTION: domain.com. 86400 IN SOA ns1.inmotionhosting.com. 2018102905 86400 7200 3600000 86400

Add +multiline to the command for more details.

dig SOA +multiline domain.com ;; ANSWER SECTION: domain.com. 86316 IN SOA ns1.inmotionhosting.com. (2018102905 ; serial 86400 ; refresh (1 day) 7200 ; retry (2 hours) 3600000 ; expire (5 weeks 6 days 16 hours) 86400 ; minimum (1 day))

Updating SOA record

Updating DNS account s forces the SOA record to update automatically within 24-48 hours. On the off chance that your SOA record hasn't updated after an ongoing change, you'll have to alter another DNS account and additionally contact Live Support to update this for you.

Update SOA in Web Host Manager.

VPS clients can force SOA record updates in WHM as root. First, I will examine on the most proficient method to sign in to your WHM (known as Web Host Manager).

- Log into WHM.
- On the left side, select Edit DNS Zone.
- Select the domain.
- Under the nameserver is the current SOA account. For instance:
- 2018102906 Serial Number
- To update the SOA account, increase the last digit pair by one. For instance,
- 2018102906 to 2018102907 or
- 2018102929 to 2018102930.
- Press Save at the bottom.

NOTE: The DNS changes made should take effect within 24-48 hours.

Procedures to Access Your Web Host Manager (WHM) from Your Domain.

WHM runs over two unique ports:

- 2086 (doesn't utilize SSL)

- 2087 (Uses SSL)

You can sign into WHM using varieties of your domain name. A portion of the choices log you in over an SSL encoded association, and others don't. For security reasons, I recommend you generally login over SSL. The following is a list of different URLs you can use to sign in to your WHM:

• https://hostname.com:2087 (SSL)

NOTE: Be sure to depose hostname.com in the model above with the hostname of your server. Also, as you are using a protected (SSL) association, if your hostname utilizes a self-marked certificate, you may get an SSL warning. It is protected to acknowledge the signal and continue.

- http://example.com/whm (non-SSL)
- whm.example.com (non-SSL)

Troubleshooting

If you can't interface with your WHM with example.com/whm, you may have a square on those ports. Utilizing the whm.example.com URL will go above the intermediary Port 80, which is the universal internet port, so you ought to have no issue connecting with whm.example.com.

Instructions To Turn A URL Into An IP Address

There are different explanations behind wanting to change over a URL to an IP address. This is a fundamental procedure, and there are different approaches to perform the transformation. In this section, I'll go over everything you have to think about the strategies accessible to change over a URL to an IP address.

What Is a URL?

A URL, otherwise called a Uniform Resource Locator, is utilized to furnish people with an easy to recall website name that can be immediately inputted into a web program. The terms URL and URI are regularly used interchangeably. Anyway, particular distinctions exist between both.

Concerning how a URL does something amazing, the enchantment exists in what's known as a DNS or Domain Name System. All URLs are related to a specific IP, and a DNS is a network that turns the URL upward and partners it to its corresponding IP address.

Definition of an IP.

An IP, which is also known as Internet Protocol address, is a one of a kind identifier given to each machine that interfaces with the internet. It permits us to handily distinguish between server A and server B. As more and more gadgets can associate with the Internet, the quantity of available IP addresses diminishes.

In website situations, think of an IP as the one valid identifier for that domain name and the URL as the epithet.

Motives to Convert a URL to IP

There can be a couple of reasons why you might need to change over a URL to an IP address.

- If a website's DNS network is down, then you may, at present, have the option to access it utilizing its IP address.

- A specific network may acknowledge IP addresses instead of an FQDN. Therefore you have to change over your URL to IP.

- You need to find out more information about the URL, including who possesses it, where it lives, or check if the domain account is right.

Approaches to Convert a URL to IP

There are numerous approaches to change over a URL to an IP address. Specific ways you may find more advantageous than others. Anyway, the following outlines two strategies you can utilize to locate the IP address of a URL.

- IP Location Finder - https://keycdn.com/gives a tool called IP Location Finder, which you can use to effortlessly find the IP address, area, and hosting supplier of a specific hostname. Essentially enter the hostname into the search field and click Lookup.

- Run a Dig - In your command-line interface, run a dig for the hostname you might want to query. To do this, basically run the following command: ***dig yourwebsite.com*** and take a gander at the answer section, which shows the A account.

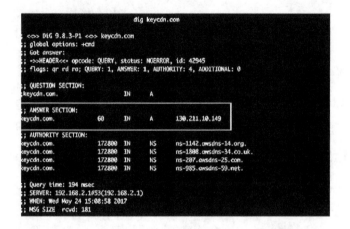

```
                         dig keycdn.com
; <<>> DiG 9.8.3-P1 <<>> keycdn.com
;; global options: +cmd
;; Got answer:
;; ->>HEADER<<- opcode: QUERY, status: NOERROR, id: 42945
;; flags: qr rd ra; QUERY: 1, ANSWER: 1, AUTHORITY: 4, ADDITIONAL: 0

;; QUESTION SECTION:
;keycdn.com.                    IN      A

;; ANSWER SECTION:
keycdn.com.             60      IN      A       130.211.10.149

;; AUTHORITY SECTION:
keycdn.com.             172800  IN      NS      ns-1142.awsdns-14.org.
keycdn.com.             172800  IN      NS      ns-1808.awsdns-34.co.uk.
keycdn.com.             172800  IN      NS      ns-207.awsdns-25.com.
keycdn.com.             172800  IN      NS      ns-985.awsdns-59.net.

;; Query time: 194 msec
;; SERVER: 192.168.2.1#53(192.168.2.1)
;; WHEN: Wed May 24 15:08:58 2017
;; MSG SIZE  rcvd: 181
```

On the off chance that you have to change over a URL to an IP address rapidly, consider using one of the two strategies referenced previously. There are other approaches to find an IP address, notwithstanding, every one of the procedures above is easy to utilize and give quick outcomes.

Now we have come to the end of the absolute beginner part of this book. I hope you were able to understand the fundamental terms discussed.

CHAPTER TWO - AMAZON WEB SERVICE

Now let's proceed to the Advanced section of this book that focuses mainly on operating the Amazon Web Service.

New companies experience various difficulties when they're trying to get by in the market. Uses are kept on the low as they endeavor to expand income. It's during this time they progress in the direction of setting up their digital footprint.

It is required to have a website when setting up a digital footprint, a website is fundamental for building the online nearness of new companies. Setting up a website becomes an issue for business owners that are, as of now, overwhelmed by business operational work. These business owners will end up spending more than would generally be appropriate properties to set up their websites because of their absence of specific information.

Business owners who plan on keeping up with technology should now get on board with the fleeting trend of cloud-based infrastructure. Adopting this technology will open chances to develop and profit by economies-of-scale.

Business owners looking to make a website come across terms like web hosting, cloud configurations, and figured out how to have services during research. While searching for web hosting, the name Amazon Web Services (AWS) frequently springs up.

Instead of business owners building the infrastructure themselves, business owners can lease the infrastructure from AWS, lowering the hindrance to the section for innovators and business people. Business owners don't need to have an on-location server center, they can depend on AWS, such as its redundancies, high accessibility, and all the things that make that infrastructure significant.

How about we start by defining the open cloud. The cloud is only an assortment of data centers. There is no ownership from the customer's point of view; the cloud supplier claims the services, and you lease each service as required. You might be thinking that the cloud is all virtual properties, yet the AWS cloud can give you open metal servers. If you need, Amazon will joyfully have your applications and databases on open metal servers facilitated in its data centers. More commonly, AWS will offer you numerous virtual servers in more than 150 different sizes and plans. Amazon is likewise glad to permit you to continue to work your on-premise data centers and exist together with cloud properties and services operating at AWS.

Another term to define before we proceed onward is Cloud computing, which is a term alluded to storing and accessing information over the internet. It doesn't store any information on the hard disk of your PC. In cloud computing, you can access data from a remote server.

Pretty much every IT solution is named with the term cloud computing or cloud nowadays. Cloud computing, or the cloud, is a similitude for the stockpile and utilization of IT properties. The IT properties in the cloud aren't legitimately apparent to the client; there are layers of reflection in the middle. The degree of deliberation offered by the cloud differs from providing virtual machines (VMs) to delivering software as a service (PaaS) in light of complex appropriated networks. Several resources are accessible on-request in gigantic amounts, and you pay for what you use.

Clouds Are Regularly Split Into Three Types:

- Open—A cloud overseen by an association and open to use by the overall population.
- Private—A cloud that virtualizes and conveys the IT infrastructure for a single association.

- Half breed— This is a blend of an open and a private cloud.

AWS is an open cloud. Cloud computing services additionally have a few groupings, such as:

- Infrastructure as a service (IaaS)— Offers significant properties like computing, storage, and networking abilities, using virtual machines, for example, Amazon EC2, Google Compute Engine, and Microsoft Azure.
- Platform as a service (PaaS)— Provides platforms to convey custom applications to the cloud, for example, AWS Elastic Beanstalk.
- Software as a service (SaaS)— has the combination of infrastructure and programming that runs in the cloud, which includes office applications such as Amazon Workspaces, and Google Apps for Work.

AWS represents Amazon Web Services. It's a worldwide market pioneer in Infrastructure as a Service (IaaS) and Platform as a Service (PaaS) industry, offering a broad scope of remote cloud services. In this section, I will cover everything about AWS for beginners so you can scale your business effortlessly.

AWS for beginners offers database storage choices, computing power, content conveyance, and networking among other functionalities to assist associations with scaling up. It permits you to choose your ideal solutions while you pay for precisely the services you expend as it were. AWS, for example, causes you to spare your currencies.

AWS offers a broad scope of remote cloud services for application improvement like examination, blockchain, Artificial Intelligence, and so forth and can help individuals and associations in the advancement and sustainable development of any application.

AWS EC2

Versatile Compute Cloud or EC2, for short, is a virtual server that causes you to run various applications on the AWS cloud infrastructure. It permits subscribers the ability to run applications in a computing situation that is fit for serving an extensive configuration of virtual systems.

With AWS, you could get instances with various property configurations of CPU, memory, storage, and networking. Each type is accessible in multiple sizes with the goal that it can take into account outstanding tasks at hand as required. Instances come from Amazon Machine Images (AMI). These machine pictures go about as a layout that configures an operating network and determines the working condition of the client. The clients can likewise design their own AMIs. You can begin deploying your bunch of servers when you have chosen your instance, alongside the operating network of your decision.

Concerning creating an EC2 instance, there are three essential ways you can use it for this reason:

- Compute Optimized – These are best reasonable for situations that require high request rates and influence industry-leading processors.
- Memory-Optimized – These instances offer the most proficient memory cost.
- Storage enhanced – These instances can access SSD storage very quickly to serve information recovery demands.

AWS EC2 service offers adaptability and an assortment of instance types for you to browse. You can customize operating networks and security settings easily. In that case, you will be liable for the provisioning limit, monitoring wellbeing, and the performance of your virtual servers.

AWS LIGHTSAIL

AWS Lightsail is an essential cloud hosting network that is comparatively more straightforward than most AWS services. While using AWS Lightsail, you can easily set up a server in only a couple of clicks. It automatically outfits your network with WordPress on AWS, Magento, and other commonly utilized web applications.

Notwithstanding its effortlessness, you should not limit or configure Lightsail as a service just for the beginners. AWS Lightsail offers an assortment of choices, for example, supporting a Windows server or a decision of Linux distros, accordingly helping experts to exploit this service bundle.

Perhaps the best bit of leeway of using AWS Lightsail is that the size of your website doesn't make a difference. This service can have your website on the AWS network easily. Moreover, with data centers in every single significant nation of the world, the clients find a good pace consistent and stable association consistently.

Furthermore, the pricing is likewise very sensible. An essential of 512MB RAM, 1 CPU center, 20GB storage, and 1TB of the monthly move is accessible for the rate of $3.50 every month as it were. On the off chance that that is too essential for you, there are other plans available. You can settle on a 4GB RAM, 2 CPU center, 80GB storage, and 4TB exchange configuration, only for $20 every month. It is incredibly affordable. It offers extraordinary additional highlights alongside an extensive network that makes this service commendable enough for thought.

AWS LAMBDA

AWS Lambda is an incredibly amazing and cost-accommodating medium that permits your business to appreciate force and adaptability consistently. This computes service permits you to run codes without managing servers.

With Lambda, you can concentrate on developing your applications without worrying about the infrastructure, for example, CPU, storage, or memory. It doesn't make a difference whether there are a couple of requests for every day or thousands every second; it executes the code when required and can scale automatically. It is an excellent platform to run applications in an AWS situation. Additionally, it permits you to extend your spending limit as you pay for what you expend.

It further permits you to transfer your code, otherwise called a Lambda work. You can design it to execute under specific conditions also. When the Lambda work is set up, it will work as much of the time as its parameters indicate.

With Lambda, you are answerable for your codes, as it deals with the compute taskforce itself, allowing you to appreciate the smooth functioning of memory, CPU, storage, and network.

The downside with Lambda is! Neither you can sign in to compute instances, nor can you customize the operating network or the language run time. These constraints permit Lambda to execute operational and administrative exercises for your sake. These exercises include monitoring fleet wellbeing, provisioning limit, deploying your code, applying security fixes, and in charge of monitoring your Lambda capacities, and so on.

The Lambda capacities which you can use on instances instead of the server-based style include:

- Application Development – It permits you to compose and execute any code without dealing with the complications of auto-scaling and infrastructure performance bottlenecks.

- Amazon S3 Cloud – It permits you to run a Lambda work when you transfer another document to an S3.

- Amazon Kinesis – It permits you to trigger Lambda works on explicit logging occasions, for example, new visitors to a website.

- CloudTrail in AWS – It permits you to dispatch occasions signed in the stack of Cloudtrail logs like enabling or disabling consents to access properties like APIs or S3 storage cans.

AWS for Businesses and Individuals?

AWS cloud storage solution offers different advantages for business visionaries. As portrayed by Amazon cloud tutorial, "AWS offers a wide scope of worldwide computing, storage, database, examination, application, and configuration services that assist associations with moving quicker, lower IT expenses, and scale applications."

THE TOP FIVE ADVANTAGES OF AWS FOR BEGINNERS

• Scalability

AWS services are affordable for businesses, everything being equal. It doesn't make a difference whether you're a startup or an entrenched online business with substantial traffic. One of the advantages of AWS for beginners is that it permits you to scale your business adequately by offering adaptability and storage solutions.

- **Commitment Free**

Another bit of leeway of using AWS cloud services is that you don't have to go under any commitment or agreement. Moreover, there's no defined minimum spend in request to utilize their service. There is every hour charge for all server-based services. It permits you to terminate the services whenever without damaging your pocket further. This element is beneficial for businesses that would prefer not to overpay for storage or services they don't utilize or require.

- **Security**

AWS offers improved security highlights like:
- 24/7 access to information specialists

- Built-in firewall which permits quite sure entrance from profoundly prohibitive to the open domain.
- IAM services for tracking client access
- Multi-factor authentication and encoded information storage abilities
-

Since information storage and security are significant components for businesses, when an association changes to a cloud service platform, it bodes well when they anticipate that security should be high class. The AWS services guarantee that the security remains hearty for businesses, everything being equal.

• Reliability

Amazon has a vast reach and a great group of specialists. It has encouraged them to construct a strong network that is reliable and steady. Numerous businesses appreciate secure and dependable associations with information that permits them to help and fabricate their information infrastructures. This is the reason

AWS cloud service is a chief decision for some businesses.

• **Flexible and Customizable**

AWS permits you to choose the programming language, operating network, and database of your liking, henceforth enabling you to build up a solution that is best for your association and your group.

The adaptability and customization of Amazon Web Services for beginners cause them to develop. Amazon's degree of customization, combined with its effortlessness and easy to understand cloud platform, convinces numerous businesses around the globe to decide on their AWS services.

· **Easy Server Management**

AWS benefits new companies, with Cloudways, it handles server related cybersecurity dangers and malware. Regular OS patching and firmware updates on the server declines vulnerabilities. On the other hand, hiring a network admin to do this physically can be costly and tedious.

While on the topic of security, the IP Whitelisting alternative permits explicit IP addresses to associate with the server. For added security, enabling the two-advance confirmation likewise shields the client from brutal force assaults.

Through the platform, the client can vertically scale the server. This is valuable when server properties get over-burden because of substantial traffic. With Cloudways, server scaling is a bit of leeway with AWS as server scaling is adaptable. Moreover, they have a pay-more only as costs arise pricing model, which drastically diminishes the expenses for business owners when vertically scaling the server. Subsequently, the business owner can benefit from the chance to amplify deals while keeping costs minimum.

Server performance enhancement is probably the most grounded advantage of using Cloudways. Their custom PHP Stack enormously improves webpage load time. Slow webpage load time altogether increases the vault rate, as visitors would prefer not to trust that a page will load.

- **Tools for Convenience**

Cloudways makes life helpful by its staging highlight. Clients can rapidly make a staging domain for their web application to test them out without affecting the live form. At the point when the progressions on the secluded web application layer are prepared, it very well may be pushed onto the live web application.

The platform likewise automates routine tasks. For instance, Cloudways Bot is a smart tool for server, application, and account-level cautions within the platform. For example, the bot can inform the client within the platform about a more current adaptation of the web application that is accessible. This regularly prompts the client to follow up on the caution or expel it completely.

Additionally, the Cloudway's platform can likewise plan preset backups. This makes server backups that are routine server maintenance obligations. Setting up the backup scheduler is convenient and straightforward.

Uses of AWS Services

Amazon Web Services are broadly utilized for different computing purposes like:

- Webpage hosting
- Application hosting/SaaS hosting
- Media Sharing (Image/Video)
- Mobile and Social Applications
- Content conveyance and Media Distribution
- Storage, backup, and calamity recuperation
- Development and test situations
- Search Engines
- Social Networking

Interacting with AWS

At the point when you interact with AWS to configure or utilize services, you make calls to the API. The API is the passage point to AWS. I will give you an outline of the tools accessible for communicating with API: *the Management Console, the command-line interface, the SDKs, and infrastructure blueprints.* We will compare the various tools, and you will figure out how to utilize every one of them while working your way through the book.

THE MANAGEMENT CONSOLE

The AWS Management Console permits you to oversee and access AWS services through a graphical user interface (GUI), which runs in each cutting edge web program (the most recent three adaptations of Google Chrome and Mozilla Firefox; Apple Safari: and Microsoft Internet Explorer). When getting started or experimenting with AWS, the Management Console is the best spot to begin. It allows you to gain a review of the various services rapidly. The Management Console is additionally a decent method to set up a cloud infrastructure for advancement and testing.

COMMAND-LINE INTERFACE

The command-line interface (CLI) permits you to oversee and access AWS services within your terminal. Since you can utilize your terminal to automate, CLI is a vital tool. You can use the terminal to make new cloud infrastructures dependent on blueprints, transfer documents to the object store, or get the details of your infrastructure's networking configuration usually. In a situation whereby you need to automate portions of your infrastructure with the assistance of a continuous integration server, the CLI is the correct tool for such activity. The CLI offers a helpful method to access the API and combine different calls into the content. You can even begin to automate your infrastructure with contents by chaining different CLI assemblies. The CLI is accessible for Windows, Mac, and Linux, and there is additionally a PowerShell variant available.

SDKs

Ensure to utilize your preferred programming language to interact with the AWS API. AWS offers SDKs for the following platforms and dialects:

- Android
- Ruby
- Browsers (JavaScript)
- Ios
- Java

SDKs are commonly used to integrate AWS services into applications. In case you're doing programming advancement and need to incorporate an AWS service like a NoSQL database or a pop-up message service, an SDK is a correct decision for the activity.

BLUEPRINTS

A blueprint is a portrayal of your network containing all properties and their conditions. Code tool evaluates your blueprint with the existing network and ascertains the means to make, update, or erase your cloud infrastructure.

Consider using blueprints if you need to control numerous or complex situations. Blue-prints will assist you with automating the configuration of your infrastructure in the cloud. You can go through them to set a network and dispatch virtual machines. Automating your infrastructure is additionally conceivable by writing your source code with the assistance of the CLI or the SDKs. However, doing so expects you to determine conditions, ensure you can update various variants of your infrastructure, and handle mistakes yourself.

CREATING AN AWS ACCOUNT

Before you can begin using AWS, it is required to have a record by the creation of an account. Your account is an enclosure for all your cloud properties. You can connect various clients to a single account if different people need access to it; as a matter of fact, your created account will have one root client. For a successful account to be created, you need the following:

- A phone number to approve your account.
- Availability of credit card to take care of bills.

SIGNING UP

The sign-up process comprises of five stages:

- Provide your login prerequisites.
- Provide your contact information.
- Provide your installment details.
- Verify your individuality.
- Choose your help plan.

1. Providing Your Login Credentials

Creating an AWS account begins with defining a one of a kind AWS account name. The AWS account name must be comprehensively one of a kind among all AWS customers. Go to AWS-in-activity $yourname and change $yourname with the name you intend to use. Adjacent to the account name, you need to indicate an email address and a password used to authenticate the root client of your AWS account. I encourage you to pick a solid secret word to forestall abuse of your account. Utilize a password consisting of in any event 20 characters. Protecting your AWS account from undesirable access is vital to maintain a strategic distance from information breaks or undesirable property utilization for your sake.

2. Providing Your Contact Information

The following stage is adding your contact information. Fill in all the necessary fields, and continue.

3. Providing Your Payment Details

The next section requests your payment information. Input your Mastercard or credit card details. There's a choice to change the money setting from USD to anyone that is more helpful for you. On the off chance that you pick this choice, the sum in USD is converted into your local cash rate toward the month's end.

4. Verifying Your Identity

The subsequent stage is to check your character. The initial step of the procedure, which is the configuration of your telephone number. After you complete the initial segment of the form, you'll get a call from AWS. A robot voice will request your PIN. However, the four-digit PIN is shown on the website, and you need to enter it using your phone. After your Personal Identity has been checked, you are prepared to continue with the final phase.

5. Choosing Your Support Plan

The last phase is to pick a help plan; For this situation, select the Basic configuration, which is free. In the event that you later make an AWS represent your business, we recommend the business strengthen the plan. You can even switch your current support plan later. when you're set, a spring up notice will show congratulating you for effectively creating an AWS account. Finally, click the Launch Management Console to sign in to your AWS for the first run through.

SIGNING IN

You currently have an AWS account and are prepared to sign in to the AWS Management Console. As referenced before, the Management Console is a web-based tool you can use to control AWS properties; it makes a large portion of the usefulness of the AWS API accessible to you. You will then visit https://console.aws.amazon.com. Enter your email address, click Next, and afterward enter your login details to sign in. After you have effectively logged in, you are forwarded to the beginning page of the Management Console.

The management console route bar comprises of seven sections:

- AWS—Start page of the Management Console, including an outline of all things considered.

- Services—Provides speedy access to all AWS services.

- Resource Groups—Allows you to get an outline of all your AWS properties.
- Custom section (Edit)— Click the alter symbol and significant intuitive services to customize the route bar.
- Your name—Lets you access billing information and your account, and furthermore lets you sign out.
- Your locale—Lets you pick your district. You will find a good pace locale. Therefore, you don't have to transform anything here at the present moment.
- Support—Gives you access to forums, documentation, and a ticket network.

In the next, you'll be introduced to a key pair so you can associate with your virtual machines.

CREATING A KEY PAIR

A key pair comprises a private key and an open key. The open key will be transferred to AWS and injected into the virtual machine. The private key is yours; it resembles your secret key, however considerably more secure. It secures your private key as though it were a secret word. It should be confidential to you, so don't lose it—hence, you won't be able to recover it. To access a Linux machine, you utilize the SSH protocol; you'll utilize a key pair for authentication instead of a secret phrase during login. While accessing a Windows machine through Remote Desktop Protocol (RDP), you'll need a key pair to unscramble the administrator secret phrase before you can sign in.

Follow these methods to construct a newfound key pair:

- Click Key Pairs in the route bar beneath Network and Security.
- Click the Create Key Pair control.
- Name the Key Pair *mykey.*

During the key-pair creation process, you download a document called mykey.pem. You should now set up that key for sometime later.

Depending on your operating network, you have to do things any other way, so please read the section that accommodates your Operating network.

LINUX AND MACOS

The main thing you have to do is change the access right of mykey.pem with the goal that no one but you can peruse the account. To do as such, run **chmod 400 mykey.pem** in the terminal. You'll find out about how to utilize your key when you have to sign in to a virtual machine in this book.

WINDOWS

Windows don't dispatch an SSH client, so you have to download the PuTTY installer for Windows and install PuTTY. PuTTY comes with a tool called PuTTYgen that can change over the mykey.pem document into a mykey.ppk account, which you'll require in the long run:

- Run the PuTTYgen application. The most significant advances are featured on the screen.

- Select RSA (or SSH-2 RSA) under Type of Key to Generate.

- Click Load.

- Because PuTTYgen shows just .ppk documents, you have to exchange the account extension of the File Name field to All Files.

- Select the mykey.pem document, and click Open.

- Confirm the exchange box.

- Substitute Key Comment to mykey.

- Click Save Private Key and then disregard the warning about saving the key without a passphrase.

Your .pem document has now been changed over to the .ppk format required by PuTTY. You'll find out about how to utilize your key when you have to sign in to a virtual machine in this book.

Make A Billing Alert

To monitor your AWS bill from the start, the compensation per-use pricing model of AWS may feel new to you, as it isn't 100% foreseeable what your bill will resemble toward the month's end. A large portion of the models in this book are secured by the Free Tier, so AWS won't charge you anything. To give you the genuine feelings of serenity expected to find out about AWS in a comfortable situation, you will make a billing alert straightaway. The billing alarm will tell you through email if your month to month AWS bill surpasses $5 USD with the goal that you can respond rapidly. In the first place, you have to enable billing alarms within your AWS account.

The following are Procedures to follow:

- The initial step is to open the AWS Management Console.
- Click your Name in the main route bar on the top.

- Select My Billing Dashboard from the spring up menu.
- Go to Preferences by using the sub route on the left side.
- Select the Receive Billing Alerts check-box.
- Click Save preference.

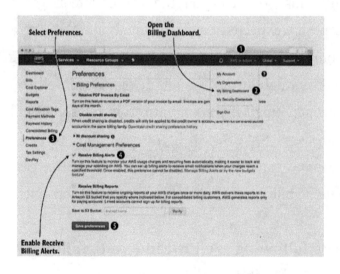

You are currently ready to make a billing alarm. Here are the means to do as such:

- Open the AWS Management Console at https://console.aws.amazon.com.

- Open Services in the route bar and select CloudWatch.
- Click the Create a billing alert link.
- Enter the limit of the total month to month charges for the billing alarm.
- Type in the email address where you need to get notices from your billing alarm if your AWS bill surpasses the limit.
- Click Create Alarm to make your billing alarm.

When you are done, open your inbox. An email from AWS containing an affirmation link shows up. Click the affirmation link to complete the configuration of your billing alarm.

Chapter Three - Developing a Virtual Infrastructure

Computing power and network availability have become a fundamental requirement for private family units, medium-sized endeavors, and huge partnerships. Presently the cloud is revolutionizing how you can access computing power. Virtual machines can be begun and stopped on-request to satisfy your computing needs within minutes. Being ready to install programming on virtual machines empowers you to execute your computing tasks without needing to purchase or lease equipment. On the off chance that you need to comprehend AWS, you need to plunge into the potential outcomes of the API working behind the scenes. You can control each service on AWS by sending requests to a REST API.

Understanding And Using REST APIs

There's a high possibility you ran over the expression "REST API" on the off chance you thought of getting information from another source on the internet, for example, Twitter or Facebook.

In any case, I will be given responses to the following questions below:

what is a REST API?

What would it be able to accomplish for you?

How would you use it?

What Is A REST API

Suppose you're trying to find a video about monster school on Youtube. You open up Youtube, and type " monster school " into an inquiry field, hit enter, and you see a list of videos about Spongebob Squared Pants. A REST API works along these lines. At the point when you look for something, and you recover a list of results from the service you're requesting from.

An API is an application programming interface. It is a lot of rules that permit projects to converse with one another. The designer makes the API on the server and permits the client to converse with it.

REST determines how the API resembles. It means "Representational State Transfer." It is a lot of rules that developers follow when they make their API. One of these guidelines expresses that you ought to have the option to get a bit of information (called a property) when you link to a particular URL.

Every URL is known as a request, while the information sent back to you is known as a reaction.

THE ANATOMY OF A REQUEST

It's critical to realize that a request is comprised of four things:

- The endpoint
- The technique
- The headers

- The information (or body)

The endpoint is the URL you demand. It follows this structure beneath:

root-endpoint/?

The root-endpoint is the starting point of the API you're requesting from. The root-endpoint of Instagram is https://api.instagram.com, while the root-endpoint Facebook API is https://api.facebook.com.

The way determines the property you're requesting. Think of it like an automatic answering machine that approaches you to press 1 for service, press 2 for another service, and 3 for one more service, etc.

You can access path simply like you can link to parts of a website. For instance, to get a list of all posts labeled under "JavaScript" on Smashing Magazine, you explore:

https://www.smashingmagazine.com/tag/javasc ript/. https://www.smashingmagazine.com/is the root-endpoint.

JSON

JSON (JavaScript Object Notation) a common format for sending and requesting information through a REST API. The response that some hosting service sends back to you is likewise formatted as JSON.

A JSON object appears as though a JavaScript Object. In JSON, every property and worth must be wrapped with twofold quotes, this way:

```
"property1": "value1",
"property2": "value2",
"property3": "value3"
```

Now let's go through the rest of the factors that make up a REQUEST.

The Method

It is the kind of request you send to the server. You can look over these five types underneath:

- GET
- POST
- PUT
- PATCH
- DELETE

These techniques give meaning to the request you're making. They are utilized to perform four potential activities: Create, Read, Update, and Delete (CRUD).

- GET: This request is utilized to get a property from a server. On the off chance that you perform a 'GET' demand, the server searches for the information you mentioned and sends it back to you. In other words, a 'GET' demand performs a 'READ' activity. This is the default demand technique.
- POST: This request is utilized to make another property on a server. If you

perform a POST demand, the server makes another section in the database and discloses to you whether the creation is effective. In other words, a 'POST' demand performs a 'Make' activity.

- PUT and PATCH: These two requests are utilized to update a property on a server. On the off chance that you perform a 'PUT' or PATCH' demand, the server updates a passage in the database and discloses to you whether the update is a success. In other words, a PUT or 'Fix' demand performs an 'UPDATE' activity.

- DELETE: This request is utilized to erase a property from a server. If you perform a DELETE demand, the server erases a passage in the database and reveals to you whether the cancellation is a success. In other words, a DELETE demand performs a DELETE activity.

The Headers

Headers are utilized to give information to both the client and the server. It very well may be utilized for some reasons, for example, authentication and providing information about the body content. You can find a list of legitimate headers on MDN's HTTP Headers Reference.

HTTP Headers are property-estimation matches that are isolated by a colon. The model beneath shows a header that orders the server to expect JSON content.

"Content-Type: application/json". Missing the opening ".

You can send HTTP headers with a twist through the - H or - header alternative. To send the above header to a hosting service API, you utilize this command:

twist - H "Content-Type: application/json" https://api.hosting service name.com

The Data (Or "Body")

The information (in some cases called "body" or "message") contains information you need to send on the server. This alternative is just utilized with POST, PUT, PATCH, or DELETE demands.

To send information through cURL, you can utilize the - d or - information alternative:

> *twist - X POST <URL> - d property1=value1*

To send numerous information fields, you can make various - d choices:

```
curl -X POST <URL> -d property1=value1 -d property2=value2
If it makes sense, you can break your request into multiple lines \ to make it easier to read:
curl -X POST <URL> \
  -d property1=value1 \
  -d property2=value2
```

Authentication

You wouldn't permit anybody to access your ledger without your consent; On a similar line of thought, developers set up measures to guarantee you perform activities just when you're approved to do. This keeps others from impersonating you.

Since POST, PUT, PATCH, and DELETE demands change the database, developers quite often put them behind an authentication divider. Now and again, a GET demand likewise requires authentication (like when you access your financial balance to check your present parity, for instance).

On the web, there are two main approaches to authenticate yourself:

- With a username and passcode
- With a token sent to you

HTTP Status Codes And Error Messages

A portion of the messages you've gotten before, "Requires authentication" and "Issues parsing JSON" are error prompt. They possibly show up when something isn't right with your request. HTTP status codes let you tell the status of the response rapidly. The range from 100+ to 500+.

All in all, the numbers adhere to the following guidelines:

- 200+ implies the request has succeeded.
- 300+ implies the request is diverted to another URL
- 400+ implies a blunder that originates from the client has happened
- 500+ implies a mistake that originates from the server has happened

You can investigate the status of a response with the verbose alternative (- v or - verbose) or the head choice (- I or - head).

For instance, on the off chance that you had a go at adding a POST demand without providing your username and secret key, you'll get a 401 status code (Unauthorized):

On the off chance that your request is invalid because your information isn't right or missing, you normally get a 401 status code (Bad Request).

Programming interface Versions

Developers renew their APIs every once in a while. Some of the time, the API can change such a lot that the developer chooses to redesign its API to another form. This occurs, and your application breaks, it's generally because you've composed code for a more established API, however your request points to the more up to date API. You can demand a particular API form in two different ways. What direction you pick relies upon how the API is composed.

These two different ways are:

- Directly in the endpoint
- In a request header

EC2

It is essential to realize that Creating virtual networks permits you to construct shut and make sure about network situations on AWS and to associate these networks with your home or corporate network.

It's amazing what you can accomplish with the computing intensity of the cell phone in your pocket or the laptop in your portable bag. In any case, if your task requires enormous computing power or high network traffic, or needs to run dependably 24/7, a virtual machine is a superior fit. With a virtual machine, you gain admittance to a cut of a physical machine situated in a server center. On AWS, virtual machines are offered by the service called Elastic Compute Cloud (EC2).

A virtual machine (VM) is a piece of a physical machine that is disconnected by programming from other VMs on the equivalent physical machine; it comprises of CPUs, memory, networking interfaces, and storage. The physical machine is known as the host machine, and the VMs running on it are called visitors. A device is answerable for isolating the visitors from one another and for scheduling requests to the equipment, by providing a virtual hard-product platform to the visitor network.

The most viral use cases for a virtual machine are as per the following:

- Hosting a web application, for example, WordPress
- Operating a venture application, for example, an ERP application
- Transforming or analyzing information, for example, encoding video account.

LAUNCHING A VIRTUAL MACHINE

It takes just a couple of clicks to install a virtual machine:

- Open the AWS Management Console at https://console.aws.amazon.com.
- Make sure you're in the right address.
- Find the EC2 service in the route bar under Services, and click it.
- Click Launch Instance to begin the wizard for launching a virtual machine.

SELECTING THE OPERATING NETWORK

The initial step is to pick an OS. In AWS, the OS comes packaged with preinstalled programming for your virtual machine; this group is called an Amazon Machine Image (AMI). Select Ubuntu Server 16.04 LTS (HVM).

The AMI is the premise your virtual machine begins from. AMIs are offered by AWS, outsider suppliers, and by the community. AWS offers the Amazon Linux AMI, which depends on Red Hat Enterprise Linux and streamlined for use with EC2. You'll likewise find famous Linux disseminations and AMIs with Microsoft Windows Server, and you can find more AMIs with preinstalled outsider programming in the AWS Marketplace.

While choosing an AMI, start by thinking about the prerequisites of the application, you need to run on the VM. Your insight and involvement in a particular operating network is another significant factor when deciding which AMI to begin with. It's additionally significant that you confide in the AMI's distributor. To be specific, I incline toward working with Amazon Linux, as it is maintained and improved by AWS.

VIRTUAL APPLIANCES ON AWS

A virtual apparatus is a picture of a virtual machine containing an OS and preconfigured programming. Since a virtual machine contains a fixed express, every time you start a VM dependent on a virtual apparatus, you'll get the very same outcome. You can duplicate virtual machines as frequently varying, so you can utilize them to eliminate the expense of installing and configuring complex heaps of programming. Virtual apparatuses are utilized by virtualization tools from VMware, Microsoft, and Oracle, and for infrastructure-as-a-service (IaaS) offerings in the cloud.

The AMI is an uncommon kind of virtual machine for use with the EC2 service. An AMI comprises of a read-just file network including the OS, additional product, and set up; it doesn't include the piece of the OS. The bit is stacked from an Amazon Kernel Image (AKI). You can likewise utilize AMIs for deploying programming on AWS.

AWS utilizes Xen, an open-source Virtual Machine Monitor. The present ages of VMs on AWS use equipment helped virtualization. The technology is called Hardware Virtual Machine (HVM). A virtual machine run by an AMI dependent on HVM utilizes a completely virtualized set of equipment and can exploit augmentations that give quick access to the underlying equipment.

Using a rendition 3.8+ part for your Linux-based VMs will give the best performance. To do as such, you should use in any event Amazon Linux 13.09, Ubuntu 14.04. In case you're starting new VMs, ensure you're using HVM pictures.

In 2017 AWS declared another age of virtualization called Nitro. Nitro combines a KVM-based Virtual Machine Monitor with customized equipment aiming to give a performance that is indistinguishable from exposed metal machines. As of now, the c5 and m5 instance types utilize Nitro.

CHOOSING THE SIZE OF YOUR VIRTUAL MACHINE

It's currently time to pick the computing power required for your virtual machine. On AWS, computing power is grouped into instance types. An instance type portrays the number of virtual CPUs and the measure of memory.

There are instance species improved for various types of utilization cases.

- T family—Cheap, moderate baseline performance with the capacity to blast to better for brief timeframes
- M family—General reason, with a fair proportion of CPU and memory
- C family—Computing improved, high CPU performance
- R family—Memory improved, with more memory than CPU power compared to M family

- D family—Storage improved, offering gigantic HDD limit
- I family—Storage improved, offering gigantic SSD limit
- X family—Extensive limit with an emphasis on memory, up to 1952 GB memory and 128 virtual centers
- F family—Accelerated computing dependent on FPGAs (field-programmable gate array)
- P, G, and CG family—Accelerated computing dependent on GPUs (designs processing units)

Instance Types And Species

The names for various instance types are organized similarly. The instance specie type all have the same attributes. AWS discharges new instance types and species every once in a while; the various forms are called generations. The instance size defines the limit of CPU, memory, storage, and networking.

The instance type t2.micro discloses to you the following:

- The instance family is called t. It combines in a small manner, modest virtual machines with low baseline CPU performance yet the capacity to blast fundamentally over baseline CPU performance for a brief timeframe.
- You're using age 2 of this instance family.
- The size is miniaturized scale, indicating that the EC2 instance is little.

Computer equipment is getting quicker and more particular, so AWS is continually introducing new instance types and species. Some of them are enhancements of existing instance species, and others are centered around explicit remaining tasks at hand. For instance, the instance family R4 was introduced in 2016. It gives instances to memory-intensive outstanding burdens and improves the R3 instance types.

Configuring Details, Storage, Firewall, and Tags

Organizing AWS Properties With Tags

Most AWS properties can be labeled. For instance, you can add tags to an EC2 instance. There are three significant use cases for property tagging:

- Use tags to channel and quest for properties.
- Analyze your AWS bill dependent on property tags.
- Restrict access to properties dependent on tags.

A firewall assist will be securing your virtual machine. The method underneath shows the settings for a default firewall allowing access utilizing SSH from anyplace.

- Select the checkbox.
- Create another security group.
- Type in ssh-just for the name and depiction of the security gathering.

- Keep the default rule allowing SSH from anyplace.
- Click Review and Launch to continue with the subsequent stage.

Reviewing Your Input And Selecting A Key Pair For SSH

You're nearly finished. The wizard should show an audit of your new virtual machine.

Ensure you picked Ubuntu Server 16.04 LTS (HVM) as the OS and t2.micro as the instance type. When every step earlier mentioned has been successfully carried out, click the Launch button. If not, return and make changes to your VM where required.

To wrap things up, the wizard requests your new virtual machine's critical.

- Pick the choice.
- Choose an Existing Key Pair,
- Select the key pair mykey,
- Click Launch Instances.

LOGGING IN WITH A KEY

Logging in to your virtual machine requires a key. You utilize a key instead of a password to authenticate yourself. Keys are substantially more secure than the latter, and using keys for SSH is enforced for VMs running on AWS.

On the off chance that you avoided the production of a key, follow these means to make an individual key:

- Open the AWS Management Console at https://console.aws.amazon.com. Find the EC2 service in the route bar under Services, and click it.
- Switch to Key Pairs employing the submenu.
- Click Create Key Pair.
- Enter *mykey* for Key Pair Name, and click Create. Your program downloads the key automatically.
- Open your terminal, and change to your download organizer.

- Linux and macOS just: change the access righto f the document mykey.pem by running ***chmod 400 mykey.pem*** in your terminal.

- Windows: Windows doesn't transport an SSH client yet, so you have to install PuTTY. PuTTY comes with a tool called PuTTYgen that can change over the mykey.pem document into a mykey.ppk account, which you'll require. Open PuTTYgen, and select SSH-2 RSA under the type of Key to Generate. Click Load. Since PuTTYgen shows just *.ppk documents, you have to switch the account augmentation of the File Name Input to All Files. Presently you can choose the mykey.pem account and click Open. Click OK in the affirmation checkbox. Change Key Comment to mykey. Click Save Private Key. Disregard the warning about saving the key without a passcode. Your .pem account is currently changed over to the .ppk format required by PuTTY.

Your virtual machine should now dispatch. Open a diagram by clicking View Instances, and hold up until the machine arrives at the Running state. To assume full responsibility for your virtual machine, you have to sign in remotely.

CONNECTING TO YOUR VIRTUAL MACHINE

Installing additional program software and running commands on your virtual machine should be possible remotely. To sign in to the virtual machine, you need to find out about its open domain name:

• Click the EC2 service in the route bar under Services, and click Instances in the submenu at left to get to the diagram of your virtual machine. Select the virtual machine from the table by clicking it. The figure underneath shows the review of your virtual machines and accessible activities.

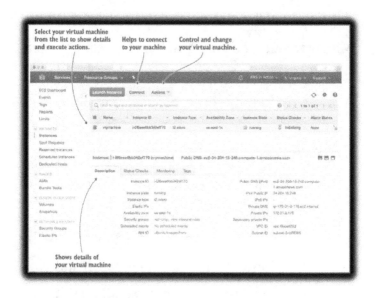

Select your virtual machine from the list to show details and execute actions.

Helps to connect to your machine

Control and change your virtual machine.

Shows details of your virtual machine

• Click Connect to open the instructions for connecting to the virtual machine.

With the open DNS and your key, you can associate with your virtual machine and Continue to the following section, depending on your OS.

LINUX AND MACOS

Open your terminal, and type *ssh - I $PathToKey/mykey.pem ubuntu@$PublicDns*, replacing $PathToKey with the path to the key account and $PublicDns with the open DNS appeared in the Connect discourse in the AWS Management Console. You'll see a security alert regarding the authenticity of the new host. Answer yes to the interface.

WINDOWS

- Find mykey.ppk account you made and double-tap to open it.
- PuTTY display ought to show up in the Windows taskbar as a symbol. If not, you may need to install or reinstall PuTTY.
- Start PuTTY. Fill in the open DNS appeared in the Connect discourse in the AWS Management Console, and click Open.
- You'll see a security alert regarding the authenticity of the new host. Answer yes and type ubuntu as the login name. Click Enter.

INSTALLING AND RUNNING PROGRAMMING

PHYSICALLY

You've presently begun a virtual machine with a Ubuntu OS. It's anything but difficult to install an additional product with the assistance of the package manager. First, we have to ensure the package manager is forward-thinking.

Run the following command to update the list of accessible packets:

$ Sudo adept get update

To begin, you'll install a tiny tool called link checker that permits you to find broken links on a website:

$ Sudo adept get install linkchecker - y

Presently you're prepared to check for links pointing to websites that never again exist. To do as such, pick a website and run the following command:

$ link checker https://...

The yield of checking the links looks something like this:

```
[...]
URL          `http://www.linux-mag.com/blogs/fableson'
Name         `Frank Ableson's Blog'
Parent URL   http://manning.com/about/blogs.html, line 92, col 27
Real URL     http://www.linux-mag.com/blogs/fableson
Check time   1.327 seconds
Modified     2015-07-22 09:49:39.000000Z
Result       Error: 404 Not Found

URL          `/catalog/dotnet'
Name         `Microsoft & .NET'
Parent URL   http://manning.com/wittig/, line 29, col 2
Real URL     http://manning.com/catalog/dotnet/
Check time   0.163 seconds
D/L time     0.146 seconds
Size         37.55KB
Info         Redirected to `http://manning.com/catalog/dotnet/'.
             235 URLs parsed.
Modified     2015-07-22 01:16:35.000000Z
Warning      [http-moved-permanent] HTTP 301 (moved permanent)
             encountered: you should update this link.
Result       Valid: 200 OK
[...]
```

MONITORING AND DEBUGGING A VIRTUAL MACHINE

If you have to find the explanation behind a blunder or why your application isn't behaving as you expect, it's imperative to approach tools that can help with monitoring and debugging. AWS gives tools that let you examine and troubleshoot your virtual machines. One methodology is to examine the virtual machine's logs.

SHOWING LOGS FROM A VIRTUAL MACHINE

On the off chance that you have to find out what your virtual machine was doing during and after startup, there is a basic solution. AWS permits you to see the EC2 instance's logs with the assistance of the Management Console (the web interface you use to begin and stop virtual machines). Follow these means to open your VM's logs:

- Open the EC2 service from the main route, and select Instances from the submenu.
- Select the running virtual machine by clicking the line in the table.
- In the Actions menu, pick Instance Settings > Get Network Log.

A window opens and shows you the network logs from your Virtual Machine that would typically be shown on a physical monitor during startup. The log contains all log messages that would be shown on the monitor of your machine if you were running it on-premises. Watch out for any log messages stating that a blunder occurred during startup. On the off chance that the mistake message isn't self-evident, you should contact the seller of the AMI, AWS Support, or post your inquiry in the AWS Developer Forums at https://forums.aws.amazon.com. This is a straightforward and effective approach to access your network logs without needing an SSH association. Note that it will take a few minutes for a log message to show up in the log watcher.

EXAMINING THE LOAD OF A VIRTUAL MACHINE

AWS can assist you with answering another inquiry such as: is your virtual machine near its maximum limit? Follow these means to open the EC2 instance's measurements:

- Open the EC2 service from the main route, and select Instances from the submenu.

- Select the running virtual machine by clicking the suitable column in the table.
- Select the Monitoring tab at the lower right.
- Click the Network In graph to jump into the details.

You'll see a diagram that shows the virtual machine's usage of incoming networking traffic. There are measurements for CPU, network, and circle use. As AWS is looking at your VM, all things considered, there is no measurement indicating the memory use. You can distribute a memory metric yourself, if necessary. The measurements are updated at regular intervals if you utilize fundamental monitoring, or consistently on the off chance that you empower point by point monitoring of your virtual machine, which costs extra.

Shutting Down A Virtual Machine

To abstain from incurring charges, you ought to consistently shut off virtual machines when you're not using them. You can utilize the following four activities to control a virtual machine's state:

• Start—You can generally begin a stopped virtual machine. If you need to make a completely new machine, you'll have to dispatch a virtual machine.

- Stop—You can generally stop a running virtual machine. A stopped virtual machine doesn't incur charges, expect for connected properties like network-appended storage. A stopped virtual machine can be begun again yet likely on an alternate host. In case you're using network-appended storage, your information continues.

- Reboot—On the off chance that you have to reboot your virtual machine, this activity is the thing that you need. You won't lose any tenacious information while rebooting a virtual machine since it remains on a similar host.

- Terminate—Terminating a virtual machine implies deleting it. You can't begin a virtual machine that you've just terminated. The virtual machine is erased, generally together with conditions like network-appended storage and open and private IP addresses. A terminated virtual machine doesn't incur charges.

NOTE: *The contrast between stopping and terminating a virtual machine is significant. You can begin a stopped virtual machine. This is unimaginable with a terminated virtual machine. On the off chance that you terminate a virtual machine, you erase it. Stopping or terminating unused virtual machines spares costs and keeps you from being shocked by a surprising bill from AWS.*

You might need to stop or terminate unused virtual machines when:

- You have propelled virtual machines to actualize a proof-of-idea. After finishing the venture, the virtual machines are never again required. Therefore, you can terminate them.

- You are using a virtual machine to test a web application. As nobody else utilizes the virtual machine, you can stop it before you knock off work, and start it back up again the following day.

• Once your clients dropped their agreement. In the wake of backing up significant information, you can terminate the virtual machines that had been utilized for your former customer.

After you terminate a virtual machine, it's never again accessible and, in the long run, vanishes from the list of virtual machines.

CHANGING THE SIZE OF A VIRTUAL MACHINE

It is constantly conceivable to change the size of a virtual machine. This is one of the advantages of using the cloud, and it enables you to scale vertically. If you need more computing power, increase the size of the EC2 instance.

In this section, you'll figure out how to change the size of a running virtual machine. To begin, follow these means to begin a little virtual machine:

- Open the AWS Management Console, and pick the EC2 service.
- Start the wizard to dispatch another virtual machine by clicking the Launch Instance.
- Select Ubuntu Server 16.04 LTS (HVM) as the AMI for your virtual machine.
- Choose the instance type t2.micro.
- Click Review and Launch to begin the virtual machine.
- Click Edit Security Groups to configure the firewall. Pick Select an Existing Security Group and select the security bunch named ssh-as it was.

- Click Review and Launch to begin the virtual machine.
- Check the list for the new virtual machine, and click Launch.
- Choose the choice.
- Choose an Existing Key Pair, select the key pair mykey, and click Launch Instances.
- Switch to the diagram of EC2 instances, and sit tight for the new virtual machine's state to change to Running.

You've presently begun an EC2 instance of type t2.micro. This is one of the littlest virtual machines accessible on AWS. Use SSH to interface with your virtual machine, as appeared in the past section, and execute.

Your virtual machine gives a single CPU center and 990 MB of memory. On the off chance that your application is having performance issues, increasing the instance size can tackle the issue. Would your application profit by additional memory? Provided that this is true, increasing the instance size will improve the application's performance too. If you need more CPUs, more memory, or more networking limit, there are numerous other sizes to browse. You can even change the virtual machine's instance family and age. To increase the size of your VM, you first need to stop it by following the below procedures:

- Open the AWS Management Console, and pick the EC2 service.
- Click Instances in the submenu to bounce to an outline of your virtual machines.
- Select your running VM from the list by clicking it.
- Choose Stop from the Actions menu.

After waiting for the virtual machine to stop, you can change the instance type by following the below procedure:

- Choose Change Instance Type from the Actions menu under Instance Settings. An exchange opens in which you can pick the new instance type for your VM.
- Select m4.large, for Instance, Type.
- Save your progressions by clicking Apply.

You've presently changed the size of your virtual machine and are prepared to begin it again. To do as such, select your virtual machine and pick Start from the Actions menu under Instance State. Your VM will begin with more CPUs, more memory, and more networking capacities. Public and private IP addresses have likewise changed. Get the new open DNS to reconnect utilizing SSH; you'll find it in the VM's Details. Use SSH to interface with your EC2 instance, and execute *cat/proc/cpuinfo and free - m* to see information about its CPU and memory.

STARTING A VIRTUAL MACHINE IN ANOTHER SERVER CENTER

AWS offers data centers everywhere throughout the world. Consider the following criteria when deciding which area to decide on your cloud infrastructure:

- Latency—Which locale offers the briefest separation between your clients and your infrastructure?

- Compliance—Are you permitted to store and administer information in that region?
- Service accessibility—AWS doesn't offer all services in all districts. Are the services you are planning to utilize accessible in the district?
- Costs—Service costs differ by area. Which locale is the most financially savvy area for your infrastructure?

Having an assumption that you have customers in the United States as well as in Chile too. Right now, you are just operating EC2 instances in N. Virginia (US). Customers from New Zealand complain about long loading occasions while accessing your website. To satisfy your Chile customers, you choose to dispatch an additional VM in Chile. Changing a server center is straightforward. The Management Console consistently shows the present server center you're working in, on the correct side of the main route menu.

ALLOCATING AN OPEN IP ADDRESS

Each Virtual Machine was associated with an open IP address automatically. In any case, each time you propelled or stopped a VM, the open IP address changed. If you need to have an application under a fixed IP address, this won't work. AWS offers a service called Elastic IPs for allocating fixed open IP addresses. You can assign an open IP address and partner it with an EC2 instance by following these means:

- Open the Management Console and go to the EC2 service.
- Choose Elastic IPs from the submenu. You'll see a diagram of open IP addresses.
- Allocate an open IP address by clicking Allocate New Address.
- Confirm by clicking on Allocate.
- Your fixed open IP address has appeared.
- Click Close to return to the outline.

Presently you can relate the open IP address with a virtual machine of your decision:

- Select your open IP address, and pick Associate Address from the Actions menu.
- Select Instance as the Resource Type.
- Enter your EC2 instance's ID in the Instance field. There is just a single virtual machine running right now, so just a single choice is accessible at that moment.
- Only one Private IP is accessible for your virtual machine. Select it.
- Click Associate to finish the procedure.

Your virtual machine is presently open through the open IP address you allotted toward the beginning of this section. Point your program to this IP address, and you should see the placeholder page.

Allocating an open IP address can be valuable if you need to ensure the endpoint to your application doesn't change, regardless of whether you need to supplant the virtual machine behind the scenes. For instance, expect that the virtual machine is running and has a related Elastic IP.

The following advances let you supplant the virtual machine with another one without changing the open IP address:

- Start another virtual machine B to supplant running virtual machine A.
- Install and start applications just as all conditions on virtual machine B.
- Disassociate the Elastic IP from virtual machine A, and partner it with virtual machine B.

Requests making use of the Elastic IP address will currently be routed to virtual machine B, with a short interruption while moving the Elastic IP. You can likewise associate multiple open IP addresses with a virtual machine by using multiple network interfaces, as depicted in the following section. This can be helpful on the off chance that it is required to have various applications running on a similar port, or on the off chance that you need to utilize a one of a kind fixed open IP address for various websites.

ADDING A NETWORK INTERFACE TO A VIRTUAL MACHINE

In addition to managing open IP addresses, you can control your virtual machine's network interfaces. It is conceivable to add multiple network interfaces to a VM and control the private and open IP addresses related to those network interfaces.

Here are some common use cases for EC2 instances with multiple network interfaces:

- Your web server needs to answer demands by using multiple TLS/SSL authentications, and you can't utilize the Server Name Indication (SNI) expansion because of inheritance clients.

- You need to make a management network isolated from the application network, and therefore your EC2 instance should be open from two networks.

- Your application requires or recommends the utilization of multiple network interfaces (for instance, network and security machines).

You utilize an additional network interface to associate a second open IP address to your EC2 instance. However, follow the means listed below to make an additional networking interface for your virtual machine:

- Open the Management Console and go to the EC2 service.
- Select Network Interfaces from the submenu.
- The default network interface of your virtual machine has appeared on the list. Note the subnet ID of the network interface.
- Click Create Network Interface.
- Enter the 2nd interface as the depiction.
- Choose the subnet you noted down.
- Leave Private IP Address vacant. A private IP will be assigned to the network interface automatically.

- Select the Security Groups that have webserver in their depiction.
- Click Yes, Create.

After the new network interface's state changes to Available, you can connect it to your virtual machine. Select the new 2nd interface network, and pick Attach from the menu. Pick the main accessible Instance ID, and click Attach. You've connected an additional networking interface to your virtual machine. Next, you'll associate an additional open IP address to the additional networking interface.

ASSOCIATING AN ADDITIONAL OPEN IP ADDRESS TO NETWORKING INTERFACE

- Open the AWS Management Console, and go to the EC2 service.
- Choose Elastic IPs from the submenu.
- Click Allocate New Address to allot another open IP address.

- Select your open IP address, and pick Associate Address from the Actions menu.
- Select the Network interface as the Resource Type.
- Enter your 2nd interface's ID in the Network Interface field.
- Select the main accessible Private IP for your network interface.
- Click Associate to finish the procedure.

We've come to the end of this section. Now let's move to the next section, which talks about AWS security.

CHAPTER FOUR - SECURING YOUR NETWORK

If security is a divider, you'll need a lot of blocks to fabricate that divider. This chapter of the book centers around the four most significant blocks to make sure about your network security on AWS:

- Installing programming updates—New security vulnerabilities are found in programming each day. Programming merchants discharge updates to fix those vulnerabilities, and you must install those updates as fast as conceivable after they're released. Otherwise, your network will be a simple one that hackers might attack.

- Restricting access to your AWS account— This becomes considerably more significant if you aren't the just one accessing your AWS account. A carriage content could, without much of a stretch, terminate all your EC2 instances instead of just the one you intended. Granting just the consents you need is critical to securing your AWS properties from unintentional or intended sad activities.

- Controlling network traffic to and from your EC2 instances—You possibly need ports to be open on the off chance that they need to be controlled. On the off chance that you run a web server, the main ports you have to open to the outside world are port 80 for HTTP traffic and 443 for HTTPS traffic.

- Creating a private network in AWS—You can make subnets that aren't reachable from the internet. And on the off chance that they're not reachable, no one can access them. Truly, no one? You'll figure out how you can gain admittance to them while preventing others from doing so.

When buying or developing applications, you ought to adhere to security standards. For instance, you have to check client input and permit just the important characters, don't spare passwords in plain content, and use TLS/SSL to encode traffic between your virtual machines and your clients.

You Should Be Acquainted With The Following Ideas Before I Could Continue:

- Subnet
- Route tables
- Access control account s (ACLs)
- Gateway
- Firewall
- Port
- Access management

WHAT IS VPC (VIRTUAL PRIVATE CLOUD), SUBNET IN AWS?

At an elevated level, you can think of a VPC in AWS as a legitimate bucket that isolates properties you make from other customers within the Amazon Cloud. It is you defining your very own network within Amazon. You can think of a

VPC like a condo where your furnishings and things are undifferentiated from databases and instances.

Amazon Virtual Private Cloud (VPC) is a coherent server center or virtual server center in Cloud. It gives a segregated section to have your machine.VPC is an assortment of the locale, Internet Gateway(IG), Route table, ACL, Security group, Subnet, and also Instances.VPC gives us a completely discrete condition where we can put our machine in a particular manner.

VPC is an assortment of the internet gateway, Router, Network ACL, EC2, Subnet, route table, and so on.

Let's have a speedy introduction about the individual.

Locale: Amazon EC2 is facilitated in multiple areas around the world. These areas are composed of Regions and Availability Zones. Every Region is a different geographic area. Every Region has multiple, separated areas known as

Availability Zones. Amazon EC2 gives you the capacity to put properties, for example, instances, and information in multiple areas.

- Internet passage is an on a level plane scaled, repetitive, and profoundly accessible VPC component that permits communication between instances in your VPC and the internet. An internet portal fills two needs, which are; to give an objective in your VPC route tables for internet-routable traffic and to perform network address interpretation (NAT) for instances that have been assigned open IPv4 addresses. Route tables contain a lot of rules, called routes, that are utilized to determine where network traffic is coordinated. Each subnet in your VPC must be related to a routing table; the table controls the routing for the subnet. A subnet must be related to each route table in turn, yet you can relate multiple subnets with a similar route table.

- A network access control list (ACL) is a discretionary layer of security for your VPC that goes about as a firewall for controlling traffic in and out of one or more subnets. You may set up network ACLs with rules like your security groups in request to add a layer of security to your VPC.VPC automatically comes with a modifiable default network ACL. It permits all inbound and outbound IPv4 traffic and, if appropriate, IPv6 traffic. One subnet can just interface with a single ACL, yet a single ACL can have multiple subnets.

- Subnetwork or subnet is a sensible subdivision of an IP network. The act of dividing a network into two or more networks is called subnetting.AWS gives two types of subnetting, which are Public, which permit the internet to access the machine, and another is private, which is avoided by the internet.

- An instance is a virtual server in the AWS cloud. With Amazon EC2, you can set up and configure the operating network and applications that sudden spike in demand for your instance.

- Amazon Virtual Private Cloud (Amazon VPC) empowers you to dispatch AWS properties into a virtual network that you've defined. This virtual network intently takes after a customary network that you'd work in your server center, with the advantages of using the adaptable infrastructure of AWS. Hence, Amazon VPC is the networking layer for Amazon EC2.

The following are the key ideas for VPCs:

- A virtual private cloud (VPC) is a virtual network devoted to your AWS account.
- A subnet is a scope of IP addresses in your VPC.

- A route table contains a lot of rules, called routes, that are utilized to determine where network traffic is coordinated.

- An internet portal is an on a level plane scaled, repetitive, and profoundly accessible VPC component that permits communication between instances in your VPC and the internet. It, therefore, forces no accessibility dangers or bandwidth constraints on your network traffic.

A VPC endpoint empowers you to secretly interface your VPC to upheld AWS services and VPC endpoint services fueled by PrivateLink without requiring an internet gateway, NAT gadget, and VPN.

Instances in your VPC don't require open IP addresses to communicate with properties in the service. Traffic between your VPC and the other service doesn't leave the Amazon network.

How Route Tables Work

Your VPC has an understood router, and you use route tables to control where network traffic is coordinated. Each subnet in your VPC must be related to a route table, which controls the routing for the (subnet route table). You can expressly relate a subnet with a specific route table. Otherwise, the subnet is verifiably connected with the main route table. A subnet must be related to each route table in turn, yet you can relate multiple subnets with the equivalent subnet route table.

You can alternatively connect a route table with internet access or virtual private access (gateway route table). This empowers you to determine routing rules for inbound traffic that enters your VPC through the gateway.

What is Internet Gateway

An internet gateway is a level plane scaled, excess, and exceptionally accessible VPC component that permits communication between instances in your VPC and the internet. It, therefore, forces no accessibility dangers or bandwidth constraints on your network traffic.

An internet gateway fills two needs: to give an objective in your VPC route tables for internet-routable traffic and to perform network address interpretation (NAT) for instances that have been assigned open IPv4 addresses.

WHAT ARE AWS FIREWALLS?

Amazon Web Services (AWS) is an open cloud service platform that underpins a wide determination of operating networks, programming dialects, structures, tools, databases, and many forms of devices. AWS utilizes a shared security model, meaning that while Amazon assumes liability for protecting the infrastructure that runs AWS services, the customer is answerable for the information and applications used by end clients. Along these lines, it is significant that customers find a way to ensure their information, applications, and networks by securing their digital property with a firewall of some kind.

The term AWS Firewall alludes to any computer security network that monitors the traffic, network, applications, or information running on the Amazon cloud. For the most part, these security networks fall into two classifications which are: Web Application Firewalls and Network Firewalls.

AWS WEB APPLICATION FIREWALL

Web Application Firewalls assume a basic job in the insurance of web-based together applications running concerning the Amazon cloud. They form the spine for cautious means against cloud-based adventures that compromise security or mischief the accessibility of uses and information. Various companies offer Web Application Firewalls on the AWS commercial center, each with their points of interest and drawbacks. While each AWS varies in technology and execution, the following are the listed commonly provided.

- **Application Security:** Protecting web applications is any Web Application Firewall's main role. An incredible Web Application Firewall (WAF) ought to have the option to ensure applications, information, APIs, and versatile application backends from common digital assaults.

- **Traffic Filtering:** Traffic filtering is one of the most commonsense and significant activities performed by a Web Application Firewall. By filtering traffic dependent on factors, for example, HTTP headers, keywords, IP addresses, and even URI strings, the Web Application Firewall can forestall unsafe interactions before they arrive at an application.

While Amazon offers its own internal firewall service, clients can regularly find more particular outsider firewalls on the AWS Marketplace better customized to their necessities.

AWS NETWORK FIREWALLS

Network Firewalls on AWS offer network insurance that compliment the application security gave by Web Application Firewalls. While there is some cover between what a Network Firewall and Web Application Firewall ensure (most remarkably information), Network Firewalls give security over the whole network border, which includes the profoundly helpless port and protocol levels. Besides, they add incredible security highlights to AWS organizations such as:

- Packet Filtering: By monitoring all incoming and outgoing packets, the firewall can direct which applications and hosts are permitted to interact with the network.

- Virtual Private Network (VPN): Many present-day firewalls offer VPN technology to permit virtual point-to-point links between two hubs through a sheltered and controlled source.

- Deep Packet Inspection (DPI): DPI is a technique that inspects the bundle's multiple headers, yet additionally, the real information substance of the packets. Along these lines, the firewall can channel protocol non-compliances, infections, spam, intrusions, or other defined criteria.

- Antivirus Inspection: Antivirus inspection checks packets for infection that makes a trip through the network to infect endpoint various devices.

- Website Filtering: Website filtering is a procedure used to check incoming web pages to check whether the page should be controlled or declined to appear by any means. Purposes behind blockage could be advertising, explicit substance, spyware, infections, and other hazardous substance.

- DNS Reputation Filtering: By filtering content against a database that accounts for the reputation and legitimacy of an IP address, firewalls can square hurtful substances with no form of problem.

Why AWS Firewalls are Important

Whether you interact with AWS all alone or not, there is a decent possibility that some close to home information of yours is stored on an AWS server worked by some association you interact with. At the point when companies have their information or applications seized by cybercriminals, it can have extensive impacts for both the company and the clients themselves. AWS Firewalls give the important assurance and security with the goal that a company's private and client information remains safe.

UNDERSTANDING SECURITY ON AWS?

Numerous associations rely upon Amazon Web Services for basic bits of their infrastructure, including storing a lot of delicate information. To protect this information, AWS gives clients a wide assortment of security services that cooperate to restrain access to approved clients. Security on Amazon Web Services (AWS) is the customizable assortment of insurances that worked to give AWS customers a sheltered space to control their accounts.

THE AWS SHARED SECURITY MODEL

In general, AWS considers it to be as ensuring the security 'of' the cloud, while customers are liable for ensuring their own security 'in' the cloud. Practically speaking, this implies customers can depend on AWS global infrastructure all in all, and the security of their information when utilized together with appropriately designed compute and storage properties. In any case, areas like substance, personality and access management, encryption, and OS configuration are the obligation of the customer.

AWS SECURITY FEATURES

- **Identity and Access Management:** a structure for managing of digital characters. Solely cloud-driven, IAM gives IT supervisors authority over clients accesses to delicate information by defining 'access jobs,' then placing clients in said jobs dependent on their security benefits.

- **Elastic Load Balancer:** manufactured and gave by AWS, an ELB can help relieve DDoS style assaults. An ELB can ensure applications by moving traffic to multiple server instances during high traffic loads.

- **AWS VPC:** a virtual private cloud service, which can aid a completely customizable and make sure about the communication between both client and server.

- **AWS Monitoring**: through tools like AWS Cloudwatch and EC2 Scripted Monitoring, the two of which fill in as completely included monitoring services. Consistent monitoring will assist with capturing any security break right away. With Amazon's monitoring services, this procedure can be automated to maintain a strategic distance from any deferrals in catching a serious security break.

- **Certificate Management**: service customers can do configuration, oversee, and send Secure Sockets Layer/Transport Layer Security (SSL/TLS) endorsements for use with AWS services. SSL/TLS endorsements are utilized to make sure about network communications and set up the nature of websites over the Internet.

- **Client-server Side Encryption Tools:** Client-side encryption alludes to encrypting information before sending it to Amazon S3. You have the following two choices for using information encryption keys. Utilize an AWS KMS-oversaw customer ace key. Server-side encryption is all about information encryption, i.e., Amazon S3 encodes your information at the object level as it composes it to disks in its data centers and unencodes it for you when you access it.

- **Hardware Security Modules:** AWS Cloud HSM is a cloud-based equipment security module (HSM) that empowers you to handily create and utilize your encryption keys on the AWS Cloud. With Cloud HSM, you can deal with your encryption keys using FIPS 140-2 Level 3 approved HSMs.

- **Web Application Firewalls:** shields your web applications from common web abuses that could influence application accessibility, compromise security, or expend over the top properties. AWS WAF gives you command over which traffic to permit or square to your web applications by defining customizable web security rules.

- **Data Encryption**: abilities accessible in AWS storage and database services, for example, EBS, S3, Glacier, Oracle RDS, SQL Server RDS, and Redshift.

- **Key Management:** including AWS Key Management Service, permits the client to decide to have AWS deal with the encryption keys or to maintain independent control of them.

- **Encrypted Message Queues**: for transmitting delicate information using encryption on the server-side.

- **Integration APIs**: integrate encryption and information security with any of the services in an AWS domain.

WHY AWS SECURITY IS SO IMPORTANT

AWS is worked to give adaptable security solutions. With close to 2,000 security controls, AWS can frequently give a lot more grounded degree of insurance, particularly for littler businesses, than could be worked in the house. A favorable position of the AWS cloud is that it allows its clients to users and innovate while being ensured a sheltered and make sure about cloud conditions. Customers just need to pay for the services they use, which diminishes forthright costs, all while maintaining at a lower cost than an on-premises workplace. AWS Security is worked around giving the client to such an extent or as a meager force as they need.

What are your duties as an AWS operator?

- Implementing access management that confines access to AWS properties like S3 and EC2 to a minimum, using AWS IAM.

- Encrypting network traffic to keep hackers from reading or manipulating information (for instance, using HTTPS).

- Configuring a firewall for your virtual network that controls incoming and outgoing traffic with security units and ACLs.

- Encrypting information. For instance, permit information encryption for your database or other storage networks.

- Managing patches for the OS and additional program software on virtual machines.

Security involves an interaction among AWS and you, then the customer. On the off chance that you carry on reasonably, you can accomplish high-security standards in the cloud.

KEEPING YOUR SOFTWARE UPDATED

Not seven days pass by without the arrival of a significant update to fix security vulnerabilities in some bit of programming or another. Some of the time, your OS is influenced, or delicate product libraries like OpenSSL. Other occasions are surrounding like Java, Apache, or an application. On the off chance that a security update is discharged, you should install it rapidly, because the endeavor may have just been discharged, or because attackers could take a gander at the source code to remake the vulnerability. You ought to have a working configuration for how to apply updates to all running virtual machines as fast as could reasonably be expected.

INSTALLING SECURITY UPDATES ON RUNNING

VIRTUAL MACHINES

What do you do if you need to install a security update of a central component on many virtual machines? You could physically sign in to all your VMs using SSH and run *yum - y - security update or yum update-to* install the security update, however, if you have numerous machines or the quantity of machines increases, One approach to automating this task is to utilize a little content that gets a list of your VMs and executes *yum* in every one of them.

The following listing unveils how this should be possible with the assistance of a Bash script.

AWS Network Manager: apply fixes in an automated way Using SSH to install security updates on the entirety of your virtual machines is challenging. It is required to have a network association just as a key for each virtual machine. Handling mistakes during applying patches is another test.

The AWS Network Manager service is an integral property while managing virtual machines. First, you install a specialist on each virtual machine. You then control your EC2 instances with the assistance of AWS SSM, for instance, by creating a vocation to fix all your EC2 instances with the most recent fix level from the AWS Management Console.

SECURING YOUR AWS ACCOUNT 'S ROOT CLIENT

It is encouraged you enable MFA for the root client of your AWS account. After MFA is initiated, a secret phrase and a brief token are expected to sign in as the root client.

Follow these means to empower Multi-factor Authentication MFA:

- Click your name in the route bar at the top of the Management Console.
- Select My Security Credentials.

- A pop up ought to show up. Click Continue to Security Credentials.

- Install an MFA application on your cell phone, one that bolsters the TOTP standard.

- Expand the Multi-factor authentication (MFA) section.

- Click Activate MFA.

- Select a virtual MFA gadget and continue with the subsequent stage.

- Follow the instructions in the wizard. Utilize the MFA application on your cell phone to check the QR code that is shown.

AWS IDENTITY AND ACCESS MANAGEMENT (IAM)

This service gives authentication and approval to the AWS API. At the point when you send a request to the AWS API, IAM confirms your personality and checks on the off chance that you are permitted to perform the activity. IAM controls who can do what in your AWS account.

For instance, is the client permitted to dispatch another virtual machine?

- An IAM client is utilized to authenticate individuals accessing your AWS account.
- An IAM group is an assortment of IAM clients.
- An IAM job is utilized to authenticate AWS properties, for instance, an EC2 instance.
- An IAM approach is utilized to define the consents for a client or task.

AUTHENTICATING AWS PROPERTIES WITH TASKS

There are different situations where an EC2 instance needs to access or oversee AWS properties. For instance, an EC2 instance may need to :

- Back up information to the item store S3.
- Terminate itself after occupation has been completed.
- Change the configuration of the private network condition in the cloud.

To have the option to access the AWS API, an EC2 instance needs to authenticate itself. You could make an IAM client with access keys and store the keys to access on an EC2 instance for authentication. Yet, doing so is a problem, particularly on the off chance that you need to pivot the entrance keys normally. Instead of using an IAM client for authentication, you should utilize an IAM task at whatever point you have to authenticate AWS properties like EC2 instances. When using an IAM task, your entrance keys are injected into your EC2 instance automatically.

CREATING A PRIVATE NETWORK IN THE CLOUD: AMAZON VIRTUAL PRIVATE CLOUD (VPC)

At the point when you make a VPC, you get your private network on AWS. Private methods you can utilize the address ranges 10.0.0.0/8, 172.16.0.0/12, or 192.168.0.0/16 to plan a network that isn't associated with the open internet. You can make sub-nets, route tables, ACLs, and portals to the internet or a VPN endpoint. Subnets permit you to isolate concerns. Make a different subnet for your database, web servers, intermediary servers, or application servers, or at whatever point you can isolate two systems. Another dependable guideline is that you ought to have in any event two subnets which are: open and private.

An open subnet has a route to the internet, whereas a private subnet doesn't. Your load balancer or web servers ought to be in the open subnet, and your database ought to dwell in the private subnet. For the purpose behind understanding how VPC functions, you'll make a VPC to have an undertaking web application. You'll likewise make a private subnet for your web servers and one open subnet for your intermediary servers. The intermediary servers absorb the greater part of the traffic by responding with the most recent adaptation of the page they have in their reserve, and they forward traffic to the private web servers. You can't access a web server straightforwardly over the internet but just through the web reserves.

Chapter Five - Storing Information In The Cloud

Storing your items: S3 and Glacier, Amazon CloudFront

Storing information comes with two difficulties: increasing volumes of information and ensuring solidity. Solving the difficulties is problematic or even unthinkable if using disks associated with a single machine. For this explanation, this part covers a progressive methodology: a conveyed information store consisting of an enormous number of machines associated with a network. Along these lines, you can store close boundless measures of information by adding additional machines to the circulated information store. And since your information is constantly stored on more than one machine, you are decreasing the danger of losing that information abruptly. You will find out about how to store pictures, reports, or some other kind of information on Amazon S3 in this section. Amazon S3 is an easy to-utilize, completely appropriate information store given by AWS. Information is overseen as the object, so the storage network is called an item store. I will tell you the best way to utilize S3 to back up your information, how to integrate S3 into your application for storing client created content, just as how to have static websites on S3. On top of that, I will introduce Amazon Glacier, a backup and archiving store. On the one hand, storing information on Amazon Glacier costs not exactly storing information on Amazon S3.

Back in the days, information was overseen in a chain of command consisting of envelopes and accounts. The document was the portrayal of the information. In an object store, information is stored as items. Each item comprises of a globally one of a kind identifier, some metadata, and the information itself.

Amazon S3

Amazon S3 is a dispersed information store, and probably the most seasoned service given by AWS. Amazon S3 is an abbreviation for Amazon Simple Storage Service. It's a normal web service that allows you to store and recover information composed as an object utilizing an API reachable over HTTPS.

Here are the uses of S3:
- Storing and delivering static website content.
- Backing up information.
- Storing organized information for an investigation.
- Storing and delivering client created content.

Getting started with S3 is by a long shot the easiest and most straightforward thing you will ever do!.

Let me tell you; you only need to Sign in to your AWS account using your IAM qualifications and select the S3 choice: This will bring up the S3 Management Dashboard as appeared in the following screen capture. You can utilize this dashboard to make, list, transfer, and erase objects from the buckets and access control rights also.

HOW TO CREATE BUCKETS

To begin with your first bucket, essentially select the Create Bucket choice from the S3 dashboard. Give an appropriate name.. Keep in mind, your bucket name should be remarkable and should begin with a lowercase character. Next, select a specific region where you might want your bucket to be made. Likewise, remember that you are not permitted to change the can's name after it has been made, so ensure you give it a right and meaningful name before you continue.

NOTE: *You are not charged for creating a bucket; you are charged uniquely for storing objects in the bucket and for transferring objects in and out of the bucket.*

You can alternatively enable the logging for your bucket also, by selecting the Set Up Logging choice. This will store point by point access logs of your bucket to an alternate one of your decision. Of course, logging of a bucket is debilitated; be that as it may, you can generally re-enable it significantly after your bucket is made. AWS won't charge you for any of the logging that it will perform; notwithstanding, it will at present charge you for the storage limit that your logs will expend on S3.

When your details are set up, select the Create choice to make your new bucket. The bucket is made within a couple of moments, and you should see the following landing page for your can.

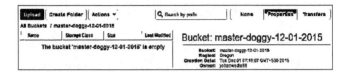

Upload | Create Folder | Actions ∨ | 🔍 Search by prefix | None | Properties | Transfers

All Buckets / master-doggy-12-01-2015

| Name | Storage Class | Size | Last Modified |

The bucket 'master-doggy-12-01-2015' is empty

Bucket: master-doggy-12-01-2015

Bucket: master-doggy-12-01-2015
Region: Oregon
Creation Dated: Tue Dec 01 07:18:57 GMT+530 2015
Owner: johanwedia86

You can even make one or more organizers in your bucket by selecting the Create Folder alternative. Envelopes are only a decent method to speak to and configure your object more adequately. You can even perform additional procedures on your bucket using this dashboard, for example, appoint authorizations, enabling logging, cross-district replication, and so on.

UPLOADING YOUR FIRST ITEM TO A BUCKET

With your bucket currently made, you can work without much of a stretch transfer any object to it. On the whole, we should investigate what an S3 object comprises:

- **Key:** This is nothing; however, one of a kind name using which you transfer objects into S3. Each item has its key,

185

which can be utilized to recognize and recover the object when vital.

- **Value:** This can be defined as a succession of bytes used to store the item's content. As examined beforehand, an object's worth can extend anyplace between zero bytes to 5 TB.

- **Version ID:** This is one more element that related to a key that can be utilized to exceptionally distinguish an object by S3. Rendition ID is similarly significant for maintaining an object's form check. Using S3, you can keep multiple variants of an object in a single can. Versioning ensures your object against coincidental overwrites just as cancellations by maintaining a different rendition number for each new item that you transfer into the bucket. Of course, versioning is debilitated on your bucket, and subsequently, your items get the rendition ID Null. You have to enable versioning on your buckets on the off chance that you wish to insure them against incidental erasures and overwrites.

- **Metadata:** These are nothing, however, basic name-esteem matches that define some information regarding a given item. There are two types of metadata given in S3: the first is the network -defined metadata, which is produced by S3 itself when an item is first transferred and it, for the most part, contains information, for example, the object's creation date, form ID, storage class, and so on. The second is client-defined metadata, which, as the name proposes, requires you as a client to give some additional information to your items when they are transferred.

- **Sub properties:** Sub properties are a lot of properties that can be related to either items or buckets. S3, as of now, supports two sub-properties with objects. The first is an Access Control List (ACL), which comprises of a list of clients and consents that are allowed access over the object. The second subresource is called torrent and is utilized to restore the torrent account related to a specific item.

- **Access control:** This gives access information of a specific item.

You can transfer questions legitimately into your buckets or within subfolders that you may have made. To begin, basically select the Upload alternative from your bucket. When the upload section pops up, −select Files and Folders exchange box.

VIEWING TRANSFERRED OBJECTS

Each transferred object in S3 is furnished with a URL that you can use to see your object using a program of your decision. The URL is in the following format:

https://s3.amazonaws.com/<BUCKET_NAME >/<OBJECT_NAME>.

You can see the URL of your object by basically selecting your item from the dashboard and the Properties alternative, as appeared. Duplicate the URL displayed against the Link characteristic tab and paste it into a web program of your decision:

NOTE: All buckets and objects in S3 are set to private, of course. You can change this default conduct by just selecting the Make Public choice from the Actions tab. This will alter your object's consents and enable everybody to see your item. You can even perform a similar activity by selecting the Permissions choice from the object's Properties tab. When the item is made open, you can see it using the URL duplicated before.

BACKING UP YOUR INFORMATION ON S3 WITH AWS CLI

Basic information should be backed up in a safe way to keep away from loss. Backing up information at an off-site area diminishes the danger of losing information in any event, during outrageous conditions like catastrophic events. In any case, where would it be a good idea for you to store your backups? S3 permits you to store any information in the form of an object. The AWS object store is an ideal fit for your backup, allowing you to pick an area for your information just as storing any measure of information with compensation for every utilization pricing model. In this section, you'll figure out how to utilize the AWS CLI to transfer information to and download information from S3. This methodology isn't constrained to off-site backups; you can utilize it in numerous other situations such as:

- Sharing documents with your associates or accomplices, particularly when working from various areas.

- Storing and retrieving relics expected to configure your virtual machines (for example, application binaries, libraries, design documents, and so on).

- Outsourcing storage ability to help the weight on local storage systems—specifically, for information that is gotten uncommonly.

First, you have to make a bucket for your information on S3. The name of the bucket must be one of a kind among all other S3 buckets, even those in other areas and those of other AWS customers. To find a one of a kind bucket name, it's helpful to utilize a prefix or addition that includes your company's name or your name. Run the following command in the terminal, replacing $yourname with your name:

```
$ aws s3 mb s3://awsinaction-$yourname
```

You utilized S3 to back up your information in the past section. If you need to diminish the expense of backup storage, you ought to consider another AWS service.

Amazon Glacier.

The cost of storing information with Glacier is about a fifth of what you pay to store information with S3. So what's the trick? S3 offers instant recovery of your information. Interestingly, you need to request your information and hold up between one minute and twelve hours before your information is accessible when working with Glacier.

Amazon Glacier is intended for archiving enormous files that you transfer once and download only sometimes. It is costly to transfer and recover a lot of little documents, so you should package little account into enormous files before storing them on Amazon Glacier. You can utilize Glacier as a standalone service open utilizing HTTPS,

CREATING A S3 BUCKET FOR THE UTILIZATION WITH GLACIER

In this section, you'll figure out how to utilize Glacier to file questions that have been stored on S3 to decrease storage costs. When in doubt, possibly move information to Glacier if the possibility you'll have to access the information later is low. For instance, assume you are storing estimation information from temperature sensors on S3. The raw data is transferred to S3 continually and prepared once every day. After the raw data has been dissected, results are stored within a database. The raw data on S3 is never again required; however, it ought to be recorded on the off chance that you have to re-run the information processing again later on. Therefore, you move the raw estimated data to Glacier following one day to minimize storage costs.

The following model aides you through storing objects on S3, moving items to Glacier, and restoring objects from Glacier.

- Open the Management Console at https://console.aws.amazon.com.
- Move to the S3 service using the main menu.
- Click the Create button.
- Type in a one of a kind name for your bucket
- Pick US East (N. Virginia) as the district for the bucket.
- Click the Next button.

- Click the Create button on the last page of the wizard.

AMAZON CLOUDFRONT

CloudFront is a Content Delivery Network (CDN) service from AWS to convey the whole web application, including streaming or interactive content, and dynamic or static content with the AWS global conveyance network of edge areas. The content request will automatically be routed to the closest edge area, so content will be conveyed with the best performance. It is enhanced to work flawlessly with other AWS services, for example, AWS S3, Amazon EC2, Amazon ELB, Route 53, and additionally with a non-AWS origin server, which will store the definitive renditions of your files. We will perceive how to use Amazon CloudFront using models as it will be more helpful for developers, and you can peruse theoretical information on the AWS website also.

Mainly, there are different choices and areas in which AWS CloudFront can work viably, for example, the following:

- Web circulation and RMTP appropriation
- Private content circulation
- Dynamic and static content
- HTTP and HTTPS protocols

AWS CLOUDFORMATION

Versatile Beanstalk, is incredible for dynamic language applications that contain close to a web level and a database level — however, imagine a scenario where your application includes additional levels to handle caching of data and other application rationale procedures, For applications like these, CloudFormation is the correct management solution. Evenly scaled alludes to the utilization of multiple computing instances, sharing the load in a single application level. Even scaling is a network for applications to help load more prominent than a single instance can handle. An alternate way to deal with solving this issue is alluded to as vertical scaling — using a better instance to help a more noteworthy burden. Vertical scaling is broadly utilized, however, it isn't the favored solution for web-scale applications, for multiple reasons, including the way that a single (exceptionally enormous) instance opens you to failure of the application instead of the repetition that littler instances give. CloudFormation activity depends on a layout — in this specific case, a JSON content archive — the format, which is the way to CloudFormation.

The following list depicts the different sections in the layout that you'll use to define your application:

- Format: Format alludes to the CloudFormation layout rendition (not the document format or some other clear term). Amazon imagines evolving the service and needs the adaptability to change the layout format to incorporate future advancements. The company is probably not going to belittle existing formats, so don't stress that your deliberately made layout will become out of date. (Note: In CloudFormation terminology, the application is alluded to as a stack, so remember this term.)

- Description: Use this (content) section to depict the layout and the application it oversees. Think of it as a Comments section, where you can give information to others as they utilize or alter your layout.

- Parameters: These qualities, which are passed into CloudFormation at runtime, can be utilized to design the application worked by the format. You may, for instance, need to run CloudFormation formats in a few Amazon areas; rather than make separate layouts for every locale, you can utilize one format and go in a parameter to define wherein district the format's application should run.

- Mappings: Here's the place you announce restrictive qualities. Think of this section, like the one in which you set a variable utilized in the layout to a specific worth. For instance, you may change the AMI ID that the layout will dispatch, in light of which locale the "area" parameter is set to.

- Resources: This area portrays the AWS properties utilized in the application and indicates the design settings. If you need the application to run all M1, large instances, place that setting there. (For a depiction of the different instance types. You can modify the setting dependent on parameters and mappings instead if you so pick.

- Outputs: These qualities are the ones you need to be returned in case of a request to portray the format. The yield may restore the name of a layout's creator or the date of creation.

FINAL REMARKS

In synopsis, Amazon's AWS cloud computing service is secure, and it is here to help you in reducing your business information infrastructure costs. Therefore, opting for AWS cloud services will demonstrate advantageous for your association and will permit you to scale your business on a leading cloud infrastructure platform exponentially.

Also, AWS security is important in such a way that AWS is worked to give adaptable security solutions. With close to 2,000 security controls, AWS can frequently give a lot more grounded degree of insurance, particularly for small businesses, than could be worked in the house. It should be understood that the favorable position of the AWS cloud is that it allows its clients to users and innovate while being ensured a sheltered and make sure about cloud conditions.

Lastly, the importance of storing data in the cloud is an important factor in such a way that you can store close, boundless measures of information by adding additional machines to the circulated information store. And since your information is constantly stored on more than one machine, you are decreasing the danger of losing that information abruptly.